Easy Learning

GCSE Foundation

SCIENCE

Revision Guide

FOR AQA A + B

Contents

CONTENTS

Periodic table

																		8
1	2			1 H hydrogen 1								3	4	5	6	7		4 He helium 2
7 Li lithium 3	9 Be beryllium 4		**Key** relative atomic mass **atomic symbol** name atomic (proton) number									11 B boron 5	12 C carbon 6	14 N nitrogen 7	16 O oxygen 8	19 F fluorine 9		20 Ne neon 10
23 Na sodium 11	24 Mg magnesium 12											27 Al aluminium 13	28 Si silicon 14	31 P phosphorus 15	32 S sulfur 16	35.5 Cl chlorine 17		40 Ar argon 18
39 K potassium 19	40 Ca calcium 20	45 Sc scandium 21	48 Ti titanium 22	51 V vanadium 23	52 Cr chromium 24	55 Mn manganese 25	56 Fe iron 26	59 Co cobalt 27	59 Ni nickel 28	63.5 Cu copper 29	65 Zn zinc 30	70 Ga gallium 31	73 Ge germanium 32	75 As arsenic 33	79 Se selenium 34	80 Br bromine 35		84 Kr krypton 36
85 Rb rubidium 37	88 Sr strontium 38	89 Y yttrium 39	91 Zr zirconium 40	93 Nb niobium 41	96 Mo molybdenum 42	[98] Tc technetium 43	101 Ru ruthenium 44	103 Rh rhodium 45	106 Pd palladium 46	108 Ag silver 47	112 Cd cadmium 48	115 In indium 49	119 Sn tin 50	122 Sb antimony 51	128 Te tellurium 52	127 I iodine 53		131 Xe xenon 54
133 Cs caesium 55	137 Ba barium 56	139 La* lanthanum 57	178 Hf hafnium 72	181 Ta tantalum 73	184 W tungsten 74	186 Re rhenium 75	190 Os osmium 76	192 Ir iridium 77	195 Pt platinum 78	197 Au gold 79	201 Hg mercury 80	204 Tl thallium 81	207 Pb lead 82	209 Bi bismuth 83	[209] Po polonium 84	[210] At astatine 85		[222] Rn radon 86
[223] Fr francium 87	[226] Ra radium 88	[227] Ac* actinium 89	[261] Rf rutherfordium 104	[262] Db dubnium 105	[266] Sg seaborgium 106	[264] Bh bohrium 107	[277] Hs hassium 108	[268] Mt meitnerium 109	[271] Ds darmstadtium 110	[272] Rg roentgenium 111								

Elements with atomic numbers 112–116 have been reported but not fully authenticated.

* The Lanthanides (atomic numbers 58–71) and the Actinides (atomic numbers 90–103) have been omitted.
Cu and Cl have not been rounded to the nearest whole number.

Coordination

The nervous system

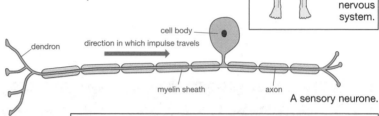

The human nervous system.

- The **nervous system** enables the body to respond to changes in the surroundings and to coordinate behaviour.

- The brain and spinal cord make up the **central nervous system**. It is connected to the body's organs by **nerves**. Nerves carry information as electrical impulses which travel to and from the spinal cord and the brain.

- In the nervous system, electrical impulses travel along cells called **neurones**.

  – The neurones that carry information from receptors to the central nervous system are called **sensory neurones**.

  – The neurones carrying information from the central nervous system to effectors are called **motor neurones**.

A sensory neurone.

A motor neurone.

G–E

D–C

Receptors

Receptors and effectors

- **Receptors** are special cells that detect stimuli. Stimuli include light, sound, changes in position, chemicals, touch, pressure, pain and temperature.

- Receptors are concentrated in the organs of our senses including the eyes, ears, nose and tongue. The skin all over our body has receptors for pressure, pain and temperature.

- Ear receptors sense a noise and send electrical impulses along nerves to the brain. The brain coordinates the response, sending impulses along nerves to neck muscles. The muscles contract and the head turns towards the noise.

- A muscle is an **effector**, an organ that does something in response to a **stimulus**.

- **Glands** are also effectors. Our salivary glands respond to a taste stimulus by secreting saliva.

G–E

Hormones

- For good health, the organs of the body must work together. Their processes are coordinated by chemicals called **hormones**. These are made by **glands** and **secreted** into the bloodstream to convey information to organs.

- Each hormone may coordinate processes in one or more **target organs**.

- Adrenaline from the adrenal gland affects the heart, breathing muscles, eyes and the digestive system. The hormones that control ovulation and menstruation target the ovaries.

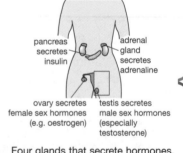

Four glands that secrete hormones.

G–E

D–C

Questions

Grades G-E
1 Name the **two** organs that make up the central nervous system.

Grades D-C
2 Name the kind of neurone that carries impulses *away* from the central nervous system.

Grades G-E
3 Give **two** examples of stimuli.

Grades D-C
4 Give an example of a hormone and its target organs.

Reflex actions

Reflex actions and synapses

- A **reflex action** is a rapid, automatic response to a stimulus, often to avoid harm, such as the blink response to an approaching object.
  - A receptor detects the stimulus.
  - An electrical impulse travels along a **sensory neurone** to the spinal cord in the central nervous system.
  - The impulse passes from the sensory neurone to a **relay neurone**, inside the spinal cord.
  - The impulse passes from the relay neurone along a **motor neurone** to an **effector** (e.g. a muscle or gland), which responds to the stimulus.

- The route of the electrical impulse in a reflex action, the **reflex arc**, is along a sensory, a relay and a motor neurone.

- The time taken is less than if the impulse travelled to the brain and back.

- The gap between the ends of two neurones is a **synapse**.

- An electrical impulse stops at the end of one neurone, which then secretes a chemical into the gap.

- The chemical diffuses across to the next neurone, which produces an electrical impulse that continues along the neurone.

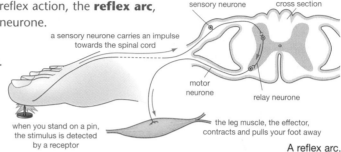

cell body of sensory neurone — spinal cord in cross section

a sensory neurone carries an impulse towards the spinal cord

motor neurone — relay neurone

when you stand on a pin, the stimulus is detected by a receptor

the leg muscle, the effector, contracts and pulls your foot away

A reflex arc.

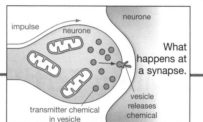

impulse — neurone — neurone

What happens at a synapse.

transmitter chemical in vesicle — vesicle releases chemical

In control

Controlling water and ions

- For cell enzymes to work effectively, cells need a steady water content, ion content and temperature. Blood sugar concentration is kept constant, to give cells a constant supply of energy.

- The kidneys excrete excess water and ions. Water is also lost through the lungs.

- Sweat contains water, ions (including sodium and chloride) and urea.

- Water evaporates, takes heat from the skin and lowers body temperature.

- During digestion, starch and sugar in food are converted to **glucose** which travels throughout the body in the blood.

- When the blood glucose level is too high, the **pancreas** secretes **insulin** into the blood.

- Insulin reaches the **liver** and makes it remove glucose from the blood and store it. Between meals, the liver can release glucose back into the blood.

How Science Works

You should be able to: evaluate the claims of manufacturers about sports drinks containing water, ions and glucose.

water from lungs in breath
water and ions in sweat
water and ions from kidneys in urine

Ways the body loses water and ions.

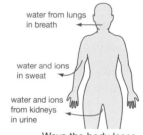

3. The water in sweat evaporates, taking heat from the skin
2. The sweat lies on the surface of the hot skin
skin
1. The sweat gland secretes sweat

How sweating cools you.

Questions

1 Name the **three** kinds of neurone involved in a reflex action.

2 Explain what happens when a nerve impulse reaches a synapse.

3 Explain why blood sugar content should be kept constant.

4 Explain how sweating cools you down.

Reproductive hormones

The menstrual cycle and hormones

- An ovary releases an egg (one cell) every 28 days.
- The lining of the **uterus** (womb) thickens, ready to receive an embryo if the egg is fertilised.
- If the egg is unfertilised, the uterus lining breaks down and is lost through the vagina in **menstruation**. This begins the sequence of events called the **menstrual cycle**.
- **Hormones** control the menstrual cycle:
  - the pituitary gland just below the brain secretes **FSH** and **LH**
  - the ovaries secrete **oestrogen**.

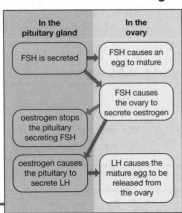

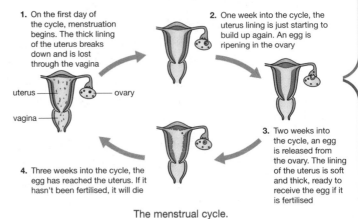

1. On the first day of the cycle, menstruation begins. The thick lining of the uterus breaks down and is lost through the vagina

2. One week into the cycle, the uterus lining is just starting to build up again. An egg is ripening in the ovary

3. Two weeks into the cycle, an egg is released from the ovary. The lining of the uterus is soft and thick, ready to receive the egg if it is fertilised

4. Three weeks into the cycle, the egg has reached the uterus. If it hasn't been fertilised, it will die

The menstrual cycle.

How hormones control the release of an egg.

Controlling fertility

Fertility treatment

- A woman who cannot produce eggs can receive **fertility drugs** which contain hormones. FSH stimulates one or more eggs to mature in the ovaries.
- In **IVF**, a woman's eggs are fertilised outside her body using her partner's sperm.
- A woman undergoing **IVF treatment** receives fertility drugs to make her ovaries produce several eggs. These are removed under anaesthetic.
- *IVF* means *in vitro* (Latin for *in glass*) *fertilisation*, since eggs were first fertilised in a glass dish.
- A fertilised egg cell divides, making a tiny ball of cells, the **embryo**. One or two embryos are transferred to the woman's uterus. They become attached to the lining and each develops into a baby.

Oral contraceptives, benefits and problems

- To avoid pregnancy, women can take oral contraceptive pills containing hormones. These include oestrogen, which stops FSH production, so eggs do not mature.
- Oral contraceptives prevent unwanted pregnancies, reduce requests for abortions and, if widely used, may slow world population growth.
- But some cultures and religions outlaw their use. They may also encourage people to have more sexual partners and increase the danger of catching sexually transmitted infections.

Questions

(Grades G-E)
1 Name the organ that secretes oestrogen.

(Grades D-C)
2 Describe **two** effects of FSH.

(Grades G-E)
3 Explain how oral contraceptives stop egg production.

(Grades D-C)
4 State **one** advantage of oral contraceptives and **one** possible disadvantage.

Diet and energy

Energy and a balanced diet

G–E

- Food provides energy for the functions of cells and for muscle movement.

- Excess energy is stored as fat, which can be used later to provide energy.

- A healthy, **balanced diet** contains the right amounts of:
  - **carbohydrates** for energy
  - **fats** for energy and making cell membranes
  - **proteins** for growth, repair and energy
  - vitamins and minerals to keep healthy
  - roughage (fibre) for digestive system function
  - water.

- People with an unbalanced diet are **malnourished**. They are too fat or too thin, or have deficiency diseases.

- The rate of chemical reactions in cells – **metabolic rate** – is generally faster in men than in women, in young than in old people, and in people with a high muscle-to-fat ratio.

- Fit, active people, who exercise or do physical work, use more energy than inactive people. Metabolic rate stays high for some time after activity.

D–C

- Metabolic rate is higher in winter than in summer.

- It can depend on the genes we inherit.

- In babies, **brown fat** cells have a very high metabolic rate to maintain body temperature.

How Science Works

You should be able to: evaluate information about the effect of food on health.

Obesity

Obesity and illnesses

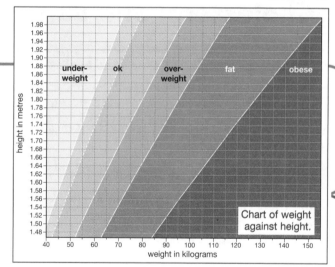

Chart of weight against height.

G–E

- People who are very overweight are described as **obese**.

- Obesity increases the chance of getting arthritis, diabetes, high blood pressure and heart disease.

- The **joints** of someone with **arthritis** are worn and painful. Many overweight people have arthritis in hip and knee joints, and have them replaced by surgery.

D–C

- People with **diabetes** cannot control their **blood glucose** level. If it rises too high, it harms cells by drawing water from them. In Type 2 diabetes, insulin fails to regulate blood glucose level.

- **Blood pressure** measures the effort the heart makes to pump blood. It is often too high in overweight people and those who eat too much salt. This can cause **heart disease** by straining the heart and damaging blood vessels.

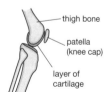

a normal knee joint has a thick cushion of smooth cartilage that allows the bones to slide easily over one another

an arthritic joint has thinner cartilage, and there are rough edges that grate against each other, causing pain and loss of mobility

A healthy knee joint and one with arthritis.

Questions

(Grades G-E)

1 List **seven** kinds of nutrient needed in a balanced diet.

(Grades D-C)

2 How does the proportion of muscle to fat in the body affect metabolic rate?

(Grades G-E)

3 Explain the meaning of the word 'obese'.

(Grades D-C)

4 Explain how arthritis makes movement difficult.

Not enough food

Eating too little or too much

- In some developing countries, drought, floods and war can prevent people from producing enough food.

- Resistance to diseases such as cholera or tuberculosis is lowered in people who do not have enough to eat. Women may have irregular or no periods.

- In rich countries, many people who over-eat try **slimming programmes** to lose weight. Eating less and exercising more would be more effective.

- Malnutrition affects small children most. Their bodies are growing fast and cannot combat infections well.

- Breastfeeding gives children a balanced diet with all the nutrients they need.

- Children between 1 and 3 years fed only on carbohydrates (e.g. maize or rice) may suffer from **kwashiorkor**. Without protein, their limbs are thin since muscle fails to develop, and their abdomen swells as fluid accumulates in the tissues. A balanced diet can bring full recovery.

A child suffering from malnutrition.

Cholesterol and salt

Cholesterol, salt and heart disease

- The bloodstream carries **cholesterol**. Too much cholesterol can block blood vessels and cause **heart disease**.

- **Saturated fats** in eggs, dairy products and meat can *raise* blood cholesterol level.

- **Monounsaturated** and **polyunsaturated fats** in plant oils can *lower* blood cholesterol.

- Cell membranes need cholesterol, and the liver makes cholesterol if the diet contains too little. The amount made seems to depend on genes.

- Blood carries cholesterol as **LDL** and **HDL** cholesterol. To be healthy, you should have more HDL than LDL in your blood.

- The body needs **salt**, but too much can cause high blood pressure and heart disease. Snacks and processed foods often contain a lot of salt.

a healthy artery has a stretchy wall and a space in the middle for blood to pass through

plaque

sometimes, a substance called plaque builds up in the wall. This is more likely to happen if you have a lot of LDL cholesterol in your blood

blood clot

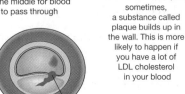

the plaque slows the blood down, and a clot may form. Or a part of the plaque may break away

How a plaque develops in an artery.

- In the blood, cholesterol mixed with protein forms tiny balls of **lipoprotein**.

- **Low density lipoproteins** (LDLs) may form **plaques** in artery walls, reducing blood flow. The blood can then clot and stop blood flow in an artery altogether. In heart muscle, when the oxygen supply from blood stops, the heart stops beating properly, and the person has a heart attack.

Questions

(Grades G-E)
1 List **three** factors that can stop people in a developing country getting enough to eat.

(Grades D-C)
2 What causes kwashiorkor?

(Grades G-E)
3 Which are best for your health – saturated fats or unsaturated fats?

(Grades D-C)
4 Which kind of cholesterol should you have more of in your blood – LDLs or HDLs?

Drugs

Drugs, addiction and dependency

Just because a drug is legal does not mean it is harmless. Legal drugs kill more people than illegal drugs.

- For thousands of years, people have used **drugs** from plants and animals. Many people use medical drugs today, to combat disease and relieve pain.

- Fewer people use other, **recreational** drugs. They change chemical processes in the brain that affect the way you feel and behave.

How Science Works

You should be able to: evaluate the different types of drugs and why some people use illegal drugs for recreation.

- Cocaine, heroin and cannabis are **illegal** drugs. Alcohol and tobacco are **legal**, used by more people, and cause more deaths.

- People who use drugs can become **addicted** to or **dependent** on a drug. Eventually, the drug damages organs such as the brain and the **liver**, the organ that destroys harmful chemicals in the body.

- Problems in a person's life can cause them to start using a drug that makes them feel better. Then they cannot manage without it. They are addicted.

- If an addicted person stops taking the drug, they can experience bad **withdrawal symptoms** for days or weeks. They need help to cure their addiction and resume a normal life.

Trialling drugs

Testing and trialling a drug

- When developing a new drug, scientists **test** its safety on tissues or animals in the laboratory to check that it is not **toxic** (poisonous).

- Then they **trial** the drug on many human volunteers to check for any side effects.

A child affected by thalidomide.

- Before general use, scientists must show that the drug does what they claim – cure an illness or ease a symptom. The whole process can take five years or more.

- **Thalidomide**, a sleeping pill, was not tested on pregnant women, and such users had babies with deformed limbs. Thalidomide was banned. Now, it is used again, to treat leprosy.

Trialling the antiviral drug zanamivir

- Zanamivir is a new drug designed to lessen flu virus symptoms. It was tested on volunteer groups – children, adolescents, young soldiers, adults, people over 65, and family members.

- In a **double-blind trial**, half of each group received the drug, and the others received a **placebo**, which contained no drug. Who received either treatment is known only after results are collected.

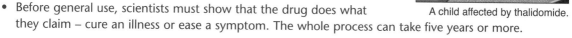

	given zanamivir	given a placebo
no. of subjects	293	295
mean age in years	19	19
no. of days until their temperature went down	2.00	2.33
no. of days until they lost all their symptoms and felt better	3.00	3.83
no. of days until they felt just as well as before they had flu	4.5	6.3
average score the volunteers gave to their experience of the five major flu symptoms	23.4	25.3

Effect of zanamivir and a placebo on soldiers suffering from flu.

Questions

Grades G-E

1 Which of these drugs are illegal: heroin, cocaine, cannabis?

Grades D-C

2 Describe what happens when a person addicted to a drug stops taking it.

Grades G-E

3 What is done with a drug before it is tried out on human volunteers?

Grades D-C

4 Explain what is meant by a 'double-blind trial'.

Illegal drugs

Cannabis, heroin and cocaine

How Science Works

You should be able to: evaluate claims made about the effect of cannabis on health and the link between cannabis and addiction to hard drugs.

G–E

- **Cannabis** can cause bronchitis and lung cancer. It is believed to increase the risk of schizophrenia, a serious mental illness. People who use cannabis often move on to take hard drugs.

- **Heroin** and **cocaine** are **hard drugs**. They are illegal and dangerously addictive. Just one use of a hard drug can cause **dependency**.

More on hard drugs

D–C

- An addicted person may inject heroin or cocaine directly into a vein to speed up its effect on the brain. If several people use the same syringe, they can spread infections such as HIV/AIDS and hepatitis.

- Obtaining drugs and the money to buy them, often through crime, takes over an addicted person's life.

- An addicted person trying to break their habit can suffer **withdrawal symptoms**. These can include insomnia, sweating, vomiting and pain all over their body.

Alcohol

Effects of alcohol

G–E

- **Alcohol** affects the nervous system. It helps people to relax and slows down their reactions.

- People drinking to excess lose their self-control and may become **unconscious**, go into a **coma**, or even die.

- Alcohol is a poison. In the long term, alcohol causes brain cells to shrink and kills the liver cells that normally try to break it down to harmless substances.

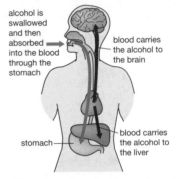

alcohol is swallowed and then absorbed into the blood through the stomach

blood carries the alcohol to the brain

stomach

blood carries the alcohol to the liver

How alcohol reaches the brain and liver.

Is alcohol an acceptable drug?

D–C

- Because alcohol has been around for thousands of years, people do not think of it as a dangerous drug. If discovered today, it would be banned; some cultures and religions do ban it.

- Drinking alcohol in moderation causes little harm, but because of excess drinking:
  - 30 000 people a year are in hospital with liver damage
  - 150 000 people a year go to hospital for other drink-related conditions
  - many people are arrested for criminal damage
  - about 4000 women and 6500 men die each year.

- Some people are **dependent** on alcohol. Like hard drugs, alcohol ruins their lives and relationships.

If you drink and drive, you increase the chance of killing and being killed.

Questions

Grades G-E

1 List **two** hard drugs.

Grades D-C

2 Why do some people who use drugs have a high risk of developing AIDS?

Grades G-E

3 Describe how alcohol affects the nervous system.

Grades D-C

4 Which organ is likely to be damaged by drinking alcohol excessively?

Tobacco

Tobacco poisons and disease

- **Nicotine** in tobacco smoke affects the brain and is **addictive**.

- **Tar** contains **carcinogens**. They cause **cancer** of the lungs and other organs.

- Cancer cells divide uncontrollably, forming a lump or **tumour**. A quarter of all smokers die of cancer.

- **Carbon monoxide** gas stops red blood cells carrying oxygen that tissues and organs need. This results in heart disease and damaged blood vessels. Babies of mothers who smoke are often small at birth.

- Tobacco smoke irritates airways and lungs, which produce **mucus** that is coughed up. Smokers get **bronchitis** when bronchi become infected, inflamed and painful. In **emphysema**, air sacs in the lungs break down, causing an oxygen shortage and breathlessness.

G–E

The link between smoking and cancer

- Richard Doll first showed the **correlation** (link) between smoking tobacco and lung cancer. Today, no-one doubts that smoking shortens life.

How Science Works

You should be able to:
- explain how the link between smoking tobacco and lung cancer gradually became accepted
- evaluate the different ways of trying to stop smoking.

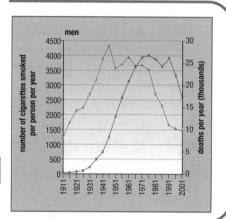

Key
○—○ cigarettes smoked per person per year
●—● deaths

The link between smoking and deaths in men.

D–C

Pathogens

Top Tip!
Do not use the word 'germs'. The correct scientific term is 'pathogens'.

Microorganisms and disease

- Microorganisms include **bacteria** and **viruses**. Bacteria are single-celled. Viruses are not made of cells and are a thousand times smaller than bacteria.

- **Pathogens** are bacteria and viruses that reproduce inside the body and cause diseases that are infectious (can be spread).

- Bacteria produce **toxins**, waste substances that the blood takes all round the body and that make you feel ill.

- Viruses invade cells, reproduce and burst the cells, destroying them, and then repeat the cycle, invading other cells.

G–E

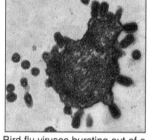

Bird flu viruses bursting out of a human cell and destroying it.

Ignaz Semmelweis

- Semmelweis was a doctor who thought that doctors' hands spread disease between patients in a hospital.

- He made doctors wash their hands, and the death rate plummeted.

How Science Works
You should be able to: relate the contribution of Semmelweis in controlling infection to solving modern problems of the spread of infection in hospitals.

D–C

Questions

(Grades G-E)
1 What is the addictive component of tobacco smoke?

(Grades D-C)
2 There is a correlation between smoking and getting lung cancer. What does 'correlation' mean?

(Grades G-E)
3 What is a pathogen?

(Grades D-C)
4 Explain why there was less disease in the hospital where Semmelweis worked, after he made doctors wash their hands.

Body defences

White blood cells, epidemics and pandemics

- White blood cells are part of the **immune system**. They protect the body against **pathogens** by:
  - **ingesting**: taking pathogens into their cytoplasm and destroying them
  - producing **antibodies** that destroy bacteria or viruses
  - producing **antitoxins** that neutralise bacterial toxins.
- Particular antibodies and antitoxins match particular pathogens.
- In an **epidemic**, many people are infected.
- A **pandemic** is a worldwide epidemic: Spanish flu killed over 50 million people in 1918.
- An epidemic takes hold when most people are not **immune** to a new pathogen.

G–E

Phagocytes, lymphocytes and antibodies

- Cells near a wound send a chemical signal to white blood cells called **phagocytes**, which gather to ingest any bacteria in the wound, in a process called **phagocytosis**.

- White blood cells called **lymphocytes** produce **antibodies**. Each is **specific** for a particular **antigen** on a pathogen. The antibody attaches to the antigen and destroys the pathogen.

- Lymphocytes also make antitoxins to destroy toxins made by bacteria.

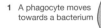
1 A phagocyte moves towards a bacterium

2 The phagocyte pushes a sleeve of cytoplasm outwards to surround the bacterium

3 The bacterium is now enclosed in a vacuole inside the cell. It is then killed and digested by enzymes

Phagocytosis.

D–C

Drugs against disease

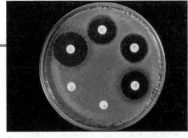
Testing antibiotics.

Painkillers and antibiotics

- Medical drugs called **painkillers** reduce pain symptoms but have no effect on the cause.

- **Antibiotics** are drugs taken into the body that kill bacterial cells but not body cells.

- In the experiment shown on the right, paper discs soaked in six different antibiotics were laid on a dish containing a layer of one type of bacterium. The bacterium failed to grow round four of the discs, showing that the discs contained antibiotics effective against the bacterium.

G–E

Antibiotics sources and antiviral drugs

- Many antibiotics, including **penicillin**, come from fungi. Some bacteria have developed **resistance** to antibiotics.

- Sources of newer antibiotics include frog skin. Others are made chemically, ensuring purity and known strength.

- Since viruses invade body cells, **antiviral** drugs are liable to kill the body cells as well as the virus.

- AZT is an antiviral drug that slows down the development of HIV/AIDS, but can cause severe side effects.

D–C

Questions

(Grades G-E)
1 What is an antibody?

(Grades D-C)
2 Which kind of blood cells produce antibodies?

(Grades G-E)
3 Explain why antibiotics are useful drugs.

(Grades D-C)
4 Explain why it is difficult to develop antiviral drugs.

Arms race

Resistance to antibiotics

G–E

- In the 1940s, **antibiotics** ended the death of millions from bacterial infections.

- By **natural selection**, bacteria develop **resistance** to antibiotics. Over-use of antibiotics increases the chance of this happening.

- **MRSA** is the strain of the bacterium *Staphylococcus aureus* that is **r**esistant to the antibiotic **m**ethicillin. MRSA infections can kill weak and very young or old people.

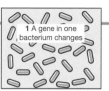
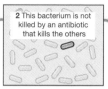
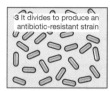

How a resistant strain of a bacterium develops.

New viral diseases

D–C

- In 1993, a new, highly infectious **virus** disease, **SARS**, first appeared in China. With international travel and without drugs to combat it, a **pandemic** (worldwide epidemic) was feared.

- In an international effort, scientists developed a **vaccine** against it, and checked for people with SARS or those who had been in contact with it. By 2003, SARS was under control.

- **Bird flu** arose in 2004, killing people in Asia. It passes from poultry to humans, but human-to-human transfer was feared if the virus mutated. World health agencies monitor this disease closely.

Vaccination

Immunisation

G–E

- Through **vaccination**, people are **immunised** (made immune) against dangerous diseases. A vaccine of harmless dead or inactivated viruses or bacteria (pathogens) is injected into the bloodstream.

- White blood cells form specific **antibodies**, which attach to **antigens** on the surface of the pathogen and destroy it. If the person is later infected with the live pathogen, the same antibodies are rapidly produced and combat the infection.

- **MMR** is a triple antiviral vaccine against measles, mumps and rubella.

How Science Works

You should be able to:

- explain how the treatment of disease has changed as a result of increased understanding of the action of antibiotics and immunity
- evaluate the consequences of mutations of bacteria and viruses in relation to epidemics and pandemics, e.g. bird flu
- evaluate the advantages and disadvantages of being vaccinated against a particular disease.

The MMR controversy

D–C

- In 1998, a team of scientists claimed in an article that the MMR vaccine caused children to develop **autism**, a personality disorder.

- Some parents refused to let their children have the MMR jab. This meant that many children were not immune to the diseases, and so could pass them on and cause an epidemic.

- Extensive studies have now proved that there is no link between MMR vaccination and autism.

year	percentage of children having MMR jab	no. of mumps cases
1996	92	94
1997	92	180
1998	91	119
1999	88	372
2000	88	703
2001	87	777
2002	84	502
2003	82	1549
2004	(no figure available)	8104

When children failed to have the MMR jab, mumps cases increased.

Questions

(Grades G-E)
1 What kind of organism is MRSA?

(Grades D-C)
2 Explain why scientists feared that SARS might become a pandemic.

(Grades G-E)
3 How does an immunisation jab make you immune to a disease?

(Grades D-C)
4 Explain why the misinformation about the MMR jab had dangerous consequences.

B1a summary

Nerves and hormones coordinate body processes and help control the body's:
- water content
- ion content
- blood sugar concentration
- temperature.

Hormones are made by glands and secreted into the bloodstream to convey information to target organs.

The menstrual cycle is controlled by FSH and LH from the pituitary and oestrogen from the ovaries.
Glucose balance is maintained by insulin produced by the pancreas.

The brain and spinal cord make up the central nervous system (CNS).
Nerves made up of neurones connect the body's organs to the CNS.

Nerves and hormones

Receptors detect stimuli and are concentrated in:
- sense organs: eyes, ears, nose and tongue
- skin: for pressure, pain and temperature.

Electrical impulses from receptors pass along nerves:
- to the brain
- to effectors (muscles and glands), which respond.

Reflex actions are rapid, automatic responses to a stimulus.
An electrical impulse travels along a reflex arc:
sensory neurone → spinal cord →
relay neurone → motor neurone → effector

Sensory neurones carry information from receptors to the CNS.
Motor neurones carry information from the CNS to effectors.
The gap between the ends of two neurones is a synapse.

A balanced diet contains the right amounts of:
- carbohydrates
- fats
- proteins
- vitamins, minerals, fibre and water.

Diet

Too much cholesterol or salt increases the chance of getting heart disease. Cholesterol levels are:
- raised by saturated fats
- lowered by monounsaturated and polyunsaturated fats.

Obesity increases the chance of getting:
- arthritis
- diabetes
- high blood pressure
- heart disease.

Medicinal drugs combat disease and relieve pain.
Recreational drugs can be:
- legal (e.g. alcohol, tobacco)
- illegal (e.g. cannabis, cocaine).

Drugs

Hard drugs such as heroin are very addictive and cause severe health problems.

Tobacco smoke contains:
- nicotine (addictive)
- carcinogens (cause cancer)
- carbon monoxide (prevents blood from carrying oxygen)
- irritants (cause bronchitis, emphysema and diseases of the heart and blood vessels).

Pathogens are microorganisms that cause infectious diseases.
Bacteria and viruses produce toxins and damage cells, which makes people feel ill.

Body defences:
white blood cells
- ingest bacteria
- produce antibodies
- produce antitoxins.

Disease

Drugs against disease:
- painkillers relieve symptoms but don't cure diseases
- antibiotics can kill bacteria, but bacteria can develop resistance to them
- immunisations and vaccinations offer protection from infectious diseases.

Hot and cold

Animals and plants are adapted to different temperatures

G–E

- Where it is hot, some animals lose heat from blood circulating through bare skin and large thin ears.

- Where it is cold, thick fur **insulates** mammals from heat loss. Small ears and compact bodies reduce **surface area**.

- Where it is hot, cacti reduce water loss with swollen stems, a thick surface, spines not leaves and a spreading, shallow root system to catch limited rainfall.

- Where it is cold, plants grow low to absorb heat from the ground warmed by sunshine and to avoid cold winds.

Examples of adaptations for extreme cold and heat

D–C

- Caribou can stand in snow at –50 °C but their legs do not freeze. Arteries and veins are close together, so arterial blood warms the blood in veins.

- Camels are adapted to a very dry desert **habitat**.

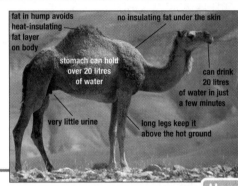

fat in hump avoids heat-insulating fat layer on body
no insulating fat under the skin
stomach can hold over 20 litres of water
can drink 20 litres of water in just a few minutes
very little urine
long legs keep it above the hot ground

How a camel is adapted to desert life.

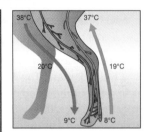

Heat exchange in a caribou's leg.

Adapt or die

How Science Works

You should be able to:
- suggest how organisms are adapted to the conditions in which they live
- suggest the factors for which organisms are competing in a given habitat
- suggest reasons for the distribution of animals or plants in a particular habitat.

Competition

G–E

- To **survive**, organisms take materials from their surroundings and from other living organisms.

- Animals **compete** for food, territory and a mate. Only those that are well **adapted** to their **environment** are likely to survive.

- Moles have a good sense of smell for finding food underground. The long mouth tube of butterflies reaches the nectar in flowers. A male peacock's tail attracts the female.

- Plants compete with each other for light and for water and nutrients from the soil.

- Plant leaves are arranged so that the greatest number receive sunlight. Their root systems are large to maximise water and nutrient intake.

Special adaptations against predators

D–C

- Sharp **thorns** protect plants from grazing animals.

- Some plants and animals produce **poisons**, and have a bright warning colour, to ensure that **predators** do not eat them. Tiny, brightly-coloured poison dart frogs in Central and South America produce poisons used by forest dwellers on their darts.

- Harmless hoverflies **mimic** (are adapted to look like) wasps that attack predators.

Questions

(Grades G-E)

1 An animal has thick fur and small ears. Is it adapted to live in a hot climate, or a cold climate?

(Grades D-C)

2 Name the habitat of a caribou, and state **one** way in which it is adapted to live there.

(Grades G-E)

3 List **three** things that animals compete for.

(Grades D-C)

4 How does the bright colour of a poison dart frog help it to survive?

Two ways to reproduce

On to the next generation

- **Genes** in the nucleus of cells carry information about characteristics from parents to offspring. Different genes control different characteristics.

- In **sexual reproduction**, an individual **inherits** half its genes from each parent. Both parents and offspring share characteristics, but different mixtures of genes make each individual differ from all others.

- **Asexual reproduction** involves only one parent. The genes of parent and offspring are identical: they are all **clones**.

G–E

Differences between sexual and asexual reproduction

- In sexual reproduction, each parent produces **sex cells** or **gametes**: females produce **eggs** and males produce **sperms**. However, in asexual reproduction, there are no gametes.

- In sexual reproduction, a female gamete fuses with a male gamete. This is **fertilisation**. However, in asexual reproduction, there is no fertilisation.

- The offspring of sexual reproduction have different mixes of genes. The parents and offspring show **variation**. However, in asexual reproduction, all the offspring are genetically identical to each other and to their parent. They are **clones**.

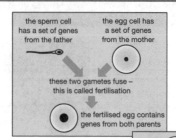

the sperm cell has a set of genes from the father — the egg cell has a set of genes from the mother — these two gametes fuse – this is called fertilisation — the fertilised egg contains genes from both parents

How an egg is fertilised in sexual reproduction.

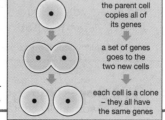

the parent cell copies all of its genes — a set of genes goes to the two new cells — each cell is a clone – they all have the same genes

Asexual reproduction – the first steps.

D–C

Genes and what they do

Genes and chromosomes

- A **gene** is a stretch of **DNA** on a **chromosome** which contains many genes. The **nucleus** of each cell contains a set of chromosomes.

- At **fertilisation**, when sex cells (gametes) fuse, the chromosomes come together. The fertilised egg has a set of genes from both parents.

- The mix of genes helps to determine an organism's characteristics.

the nucleus contains chromosomes — chromosome - a long chain of genes — DNA contains genes — gene: these control the way we grow and develop

Looking at the detail – from genes to nucleus.

G–E

- Other characteristics are not caused by genes. They result from events in an organism's life. Examples are tattoos in humans or a tree losing a branch.

- Some characteristics, such as an unattached earlobe, result from just one gene.

- The monk Gregor Mendel was the pioneer of inheritance studies 150 years ago. He bred generations of pea plants, and by good chance chose to observe characteristics that were each determined by one gene on a different chromosome.

- From his observations, he correctly proposed how genes were passed from parents to offspring.

D–C

Questions

Grades G-E

1 The offspring of asexual reproduction are 'clones'. What does this mean?

Grades D-C

2 Describe **one** difference between sexual reproduction and asexual reproduction.

Grades G-E

3 Where are genes found?

Grades D-C

4 Explain how the shape of your earlobes is determined.

Cuttings

Growing plants from cuttings

* Plants can be produced asexually by taking **cuttings**. **Hormone rooting powder** speeds up new root growth.

* The new plants and the parents are **genetically identical**, so they are clones.

* Cuttings are best taken of the **stem**, **leaf** or **root** of a young, actively growing plant:

  – all new plants are genetically identical to the original plant and to each other

  – all are likely to grow at the same rate

  – taking cuttings is easy, cheap and guarantees that all plants show the same characteristics.

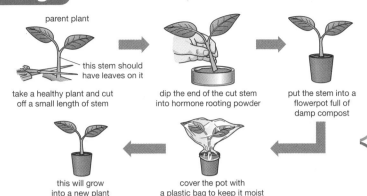

parent plant

this stem should have leaves on it

take a healthy plant and cut off a small length of stem

dip the end of the cut stem into hormone rooting powder

put the stem into a flowerpot full of damp compost

this will grow into a new plant

cover the pot with a plastic bag to keep it moist

How to take cuttings.

Clones

Tissue culture and embryo transplants

* In plant **tissue culture**, small groups of cells from a plant grow on a jelly containing all the nutrients a plant requires. The cells divide and grow into new plants, all with the same genes: they are **clones**.

* In animal **embryo transplantation**:

  – a fertilised egg (a **zygote**) is taken

  – it divides to form a ball of cells, all genetically identical

  – the cells are separated before they become **specialised**

  – each cell divides to form a ball of new cells (an **embryo**)

  – each identical embryo is **transplanted** into a host mother where it develops into a new individual.

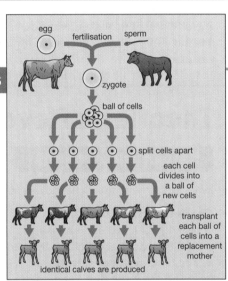

egg   fertilisation   sperm

zygote

ball of cells

split cells apart

each cell divides into a ball of new cells

transplant each ball of cells into a replacement mother

identical calves are produced

Embryo splitting and transplants – many offspring from one zygote.

Adult cell cloning and replacing damaged skin

* In **adult cell cloning**, the nucleus of an egg cell is removed and replaced with the nucleus of an adult **body cell**. The egg cell develops into an individual which is a clone of the body cell donor.

* New skin can be grown to form **skin grafts**. A victim of burns often needs skin grafts, but may not have enough undamaged skin for them. Instead, a few healthy skin cells from the victim are **cultured** in a **growth medium**, to produce large sheets of skin which can then be transferred to burnt areas by plastic surgery.

Skin ready for plastic surgery.

Questions

(Grades G-E)

1 What is a cutting?

(Grades D-C)

2 List **two** advantages of growing new plants from cuttings.

(Grades G-E)

3 Name the process by which a whole plant is grown from a little group of cells.

(Grades D-C)

4 Describe how adult cell cloning is done.

Genetic engineering

GM soya beans and insulin

- **Genetic engineering** is the technique of transferring a gene from the chromosome of one organism to another, which is then **genetically modified** (**GM**).

- A gene for a desired characteristic from one organism is inserted into embryos of a plant. When they have grown into adult plants, they may show the desired characteristic.

- Soya beans are a rich protein source. Soya plants compete with weeds. The **herbicide** that kills weeds also kills soya. A gene for **resistance** to the herbicide was inserted into a chromosome in soya, making it a **GM crop**.

How Science Works

You should be able to:
- interpret information about cloning techniques and genetic engineering techniques
- make informed judgements about the economic, social and ethical issues concerning cloning and genetic engineering, including GM crops.

- People with **diabetes** lack the hormone **insulin** to regulate their blood sugar levels.

- By genetic engineering, the gene for producing insulin was cut from human DNA by a special **enzyme**, and inserted into bacteria which provide insulin on an industrial scale.

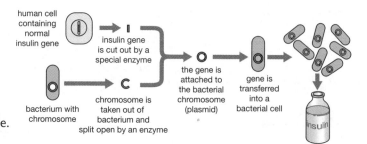

human cell containing normal insulin gene → insulin gene is cut out by a special enzyme

bacterium with chromosome → chromosome is taken out of bacterium and split open by an enzyme

the gene is attached to the bacterial chromosome (plasmid)

gene is transferred into a bacterial cell

the gene makes the bacterial cell produce insulin

insulin

The genetic engineering of bacteria to produce insulin.

Theories of evolution

How Science Works

You should be able to:
- identify the differences between Darwin's theory of evolution and conflicting theories
- suggest reasons for the different theories
- suggest reasons why Darwin's theory of natural selection was only gradually accepted.

Darwin and evolution

- Evolution is the gradual change in a species' characteristics over time.

- **Charles Darwin**'s book *On the Origin of Species* was based on his knowledge of fossils and observations of living organisms. He suggested that **species** gradually changed from one form to another as a result of **natural selection**. Darwin's theory replaced others, including the belief that all organisms were created by God in the same form as they are today, and Lamarck's idea that characteristics **acquired** in an organism's lifetime can be passed on to the next generation.

- Darwin collected finches from several different islands in the Galapagos. He noticed that the different kinds of finches all looked quite similar to each other, but had different kinds of beaks.

- He suggested that they had all evolved from one kind of finch that had arrived on the islands from the mainland. Different kinds of beaks had gradually evolved as adaptations to eat different kinds of food.

The characteristics of finches adapted to eat different kinds of food.

type of finch		food sources
ground finch		eat seeds which they crush with their strong beaks
cactus finch		long, slender beaks suck up nectar from cactus flowers
vegetarian tree finch		curved beaks, like a parrot, eat buds and fruit
insectivorous finch		stubby beaks to eat beetles and other insects
woodpecker finch		use a cactus spine to prise small insects out from cracks in tree bark

Questions

(Grades G-E)
1 Explain what is meant by a 'GM organism'.

(Grades D-C)
2 What kind of GM organism is used to produce insulin?

(Grades G-E)
3 Name **two** people who put forward different theories of evolution.

(Grades D-C)
4 Why do the Galapagos finches have different kinds of beaks?

Natural selection

Genes and survival

- Organisms compete for survival. Those with characteristics best suited to their surroundings are more likely to survive, while others are more likely to die. This is **natural selection**.

- Different characteristics are controlled by different **genes**. Occasionally a gene undergoes a change or **mutation** and so the characteristic changes. If the changed characteristic increases an organism's survival and breeding chances, it is more likely to pass to later generations.

- This can cause a **species** to change, sometimes rapidly.

Top Tip!

Remember that organisms like the moths do not change on purpose – it happens through natural selection.

Natural selection in action

- Peppered moths are usually pale with dark speckles. They are **camouflaged** from predator birds.

- In cities, the industrial revolution blackened buildings and trees with soot. Dark moths – a **variation** in the species – were better camouflaged. The dark moths survived, while birds ate the more visible, pale moths. Dark moths were being favoured by **natural selection**.

- Now, with cleaner air, most peppered moths are pale again.

Can you see the light-coloured moth in this photograph?

Fossils and evolution

How Science Works

You should be able to: suggest reasons why scientists cannot be certain about how life began on Earth.

Fossils and the origins of life

- A **fossil** forms when a dead organism is buried in **sediment**, more sediment piles on top, the pressure (and temperature) increases and the organism turns to stone.

- Comparing **fossil** remains with present-day plants and animals shows us that, since life began, some organisms have changed (**evolved**) a lot, and some only a little.

- Fossils and the **theory of evolution** support the idea that all present-day species evolved from the first **life-forms** that existed about 3.5 billion years ago.

- We can learn about the evolutionary relationships and ecological relationships between organisms by looking at similarities and differences between them.

What fossils tell us

- All present-day species have evolved from simpler life-forms.

- There are gaps in the fossil record for most species.

The fossil record for the horse shows much change in 60 million years.

years ago	60 million	40 million	10 million	modern horse
shape, size and front leg bones	height: 0.4 m	height: 0.6 m	height: 1.0 m	height: 1.6 m
how it lived	on soft ground near water; its feet could support its weight without sinking into the mud	in drier conditions; needed to be able to run away from predators	in very dry conditions; it was a fast runner	on grassland; it is a fast runner

Questions

(Grades G-E)

1 What is the name for a change in a gene?

(Grades D-C)

2 Explain why the black variety of the peppered moth became more common in the industrial revolution.

(Grades G-E)

3 How does a fossil form?

(Grades D-C)

4 What do fossils tell us about the habitats in which the very first horses lived?

Extinction

Causes of extinction

- A species that cannot adapt to change may become **extinct**; if:
  - a change occurs to its **physical** (non-living) **environment**
  - a new **predator** arrives
  - a new **disease** appears
  - a more successful **competitor** moves into its **habitat**.
- Dodos became extinct on Mauritius when rats arriving with people ate dodo eggs.
- The 30 m high giant club-moss became extinct 300 million years ago when the climate cooled and its swampy habitat dried out.

The dodo became extinct
300 years ago.

- When there is an **environmental** change, such as temperature change, organisms that adapt or move away will survive; others will die out.
- Building on green sites destroys habitats so species become locally extinct.
- A **predator** newly arrived in a habitat may kill and eat all individuals of a species not adapted to escape it.
- A species may not be adapted to withstand a new **disease**.
- Successful **competitors** for food may survive while less successful competitors become extinct.
- Hunting and habitat destruction by humans has caused many extinctions.

This is the last St Helena
olive tree, a species made
extinct by a fungal disease.

More people, more problems

Human populations and population growth

- By 2035, the world's **population** may reach 10 billion, straining the supply of **raw materials** and affecting the **environment**.
- As living standards and the number of people rise, **non-renewable energy resources** such as **crude oil** and **coal**, and minerals such as **metal ores** and **rocks**, are rapidly used up.
- **Waste** and **pollution** will increase if they are not regulated by laws.
- Until the 1800s, the human population was small. Disease, bad housing and insufficient food limited the length of people's lives.

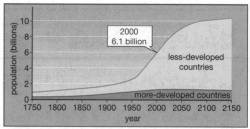

Human population growth – will it rise as predicted?

- Then, industrial **development** in western countries brought wealth and improved conditions. People lived longer with better food, medical care and safe water supplies.
- There is now a **population explosion** in many developing countries, while populations in developed countries are roughly stable.
- Bigger populations require more land for buildings, roads and food production, so the habitats of plants and other animals are destroyed.

Questions

(Grades G-E)
1 Explain why the dodo became extinct.

(Grades D-C)
2 Explain how a new competitor could cause a species to become extinct.

(Grades G-E)
3 Why may pollution by waste increase if we do not regulate it?

(Grades D-C)
4 Look at the graph above. What is happening now to the *rate* of population growth?

Land use

Loss of the natural environment

- Expanding populations lead to natural landscapes being lost:
  - people need more **buildings** for homes, workplaces and shops
  - **building materials** are cut from quarries and dredged from sand and shingle beds
  - more **food** is needed, so natural habitats become farmland
  - **waste** increases and is disposed of in landfill sites (or in incinerators).

Intensive farming

- To maximise food output and minimise prices, **intensive farming** methods are used:
  - herbicides and pesticides (toxic chemicals) and fertilisers increase the production of crops and livestock
  - hedgerows are removed to enlarge fields and make it possible to use combine harvesters
  - hens and other livestock are reared in vast battery sheds.
- Populations of wild plants and animals are wiped out by these processes.

Intensive farming replaces small fields of different crops with a huge field of one crop.

Pollution

Top Tip!

Properly treated sewage is absolutely safe – it is untreated sewage that causes problems.

How water is polluted

- **Pollution** occurs when damaging materials are put into the environment.
  - **Toxic chemicals**, for example in an oil spill, can poison living organisms.

name of disease	its symptoms
gastroenteritis	stomach cramps, diarrhoea and vomiting
Weil's disease	liver and kidney damage, it can be fatal
allergic alveolitis	fever, breathlessness, cough

Diseases you can get from untreated sewage.

  - **Sewage** contains human waste and rubbish including toilet paper, industrial waste and microorganisms. If not properly treated at sewage works and suitably disposed of, it can cause human diseases and pollute rivers, killing aquatic organisms.
  - Excess **fertiliser** applied to farmland can reach streams and rivers. Surface water plants grow rapidly and blanket the waterways. Light for photosynthesis fails to reach the plants below and they die. Bacteria feed on the dead plants, respire and use up oxygen, killing organisms such as fish and snails.

Water pollution: effects on humans and other animals

- Because **sewage** is a danger to health, all who work in sewers, toilets and water treatment plants and who transport sewage sludge, should wear protective clothing.

- In one or two very rare cases, very high concentrations of nitrates from **artificial fertilisers** reached the drinking water supply and caused blue baby syndrome, when the blood of babies cannot transport sufficient oxygen.

- The artificial pesticide **DDT** was widely used 50 years ago, especially against mosquitoes and other parasites. It did not break down and passed up the food chain to reach lethal concentrations in higher animals. Now its use is banned.

grebes (1600 p.p.m. of DDT)

fish (250 p.p.m. of DDT)

(5 p.p.m. of DDT)

plankton

(0.02 p.p.m. of DDT)

water

How DDT is taken up the food chain.

Questions

Grades G-E

1 List **four** ways in which humans reduce the amount of land available for other organisms.

Grades D-C

2 What is meant by 'intensive farming'?

Grades G-E

3 List **three** causes of land and water pollution.

Grades D-C

4 Explain why DDT killed many predatory birds.

Air pollution

Pollution by smoke, smog and sulfur dioxide

- Power stations and traffic produce **smoke**, tiny particles suspended in the air.

- Smoke particles plus water droplets in fog form **smog**.

- Burning fossil fuels, especially coal in power stations, forms **sulfur dioxide**, an acidic gas. Sulfur dioxide helps form **acid rain**.

- Smoke particles and sulfur dioxide damage the lungs.

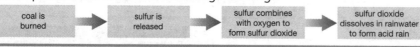

London, 1952, when over 3500 people died because of the smog.

coal is burned	→	sulfur is released	→	sulfur combines with oxygen to form sulfur dioxide	→	sulfur dioxide dissolves in rainwater to form acid rain

How acid rain is formed.

G–E

Effects of air pollution, and remedies

- **Sulfur dioxide**:
  - causes bronchitis, particularly in the elderly and young, and triggers asthma attacks
  - turns plant leaves yellow, so they cannot photosynthesise
  - **reacts** with limestone on buildings.

- **Smoke**:
  - causes coughing, lung damage and asthma attacks
  - settles as black **soot** on buildings.

- To reduce pollution by particles and harmful chemicals:
  - many factories fit filters to clean the smoke they emit
  - power stations use gas scrubbers
  - cars are fitted with catalytic converters.

Smoke can turn buildings black.

D–C

What causes acid rain?

Top Tip!

Do not confuse acid rain with the greenhouse effect! One is caused by sulfur dioxide, and the other by carbon dioxide.

More about acid rain

- Several acidic gases are present in polluted air.
  - Power stations that generate electricity by burning **fossil fuels** produce sulfur dioxide. This gas reacts with oxygen and water in the air to produce **sulfuric acid**.
  - Oxides of nitrogen in traffic exhaust form **nitric acid**.

- Rainwater is naturally weakly acidic: **carbon dioxide + water → carbonic acid**

- The acids dissolved in rainwater form **acid rain** that damages plants and **erodes** stone.

G–E

Why trees and lakes die and stone erodes

- **Trees** die because acid rain:
  - strips the protective waxy surface from leaves, the cells beneath shrivel and so cannot photosynthesise
  - washes useful nutrients and minerals out of the soil, and releases aluminium which is toxic to tree roots.

- Acid rain causes **lakes** to become so acidic that fish and other creatures are killed. **Lime**, a base, is added to acidic lakes to neutralise the acid.

- **Limestone** and **sandstone** statues and buildings are eroded because acid rain reacts with the calcium carbonate they contain and washes away their surfaces.

D–C

Questions

Grades G-E

1 Name a gas that helps to cause acid rain.

Grades D-C

2 How can sulfur dioxide affect people's health?

Grades G-E

3 List **three** acids that may be present in acid rain.

Grades D-C

4 How does acid rain harm limestone buildings?

Pollution indicators

Organisms that indicate pollution levels

G–E

- **Lichens** are often used as indicators of sulfur dioxide **air pollution**.
- **Invertebrate animals** are often used as indicators of **water pollution**.

Investigating the effects of sewage

D–C

- This table shows the results of a study of river pollution by **sewage**, using **invertebrates** as pollution level indicators.

distance downstream from sewage entry point (m)	what the water is like	invertebrates found	oxygen levels
sewage enters here 0–10	dark and cloudy; very smelly	*Chironomus* larva; rat-tailed maggot	falling quickly
10–100	cloudy; bad smell	tubifex worm (sludge worm); mosquito larva	very low
100–200	beginning to clear; slight smell	flatworm; caddis fly larva	gradually rising
200+	clear	freshwater shrimp; mayfly larva; stonefly larva	back to normal

The effects of untreated sewage in a river.

Deforestation

Deforestation and its effects

G–E

- **Deforestation** is the permanent clearing of forests, particularly of **tropical rainforests**.
- Trees are cut for **timber** or cleared (often burned) to make way for **agriculture**.
- **Photosynthesis** uses **carbon dioxide** from the air to produce carbon compounds in plant tissue.
- Much more carbon is 'locked up' in a tree trunk than in crops occupying the same area. When wood burns, carbon dioxide returns to the air.
- Deforestation reduces the rate of carbon dioxide removal from the air.
- Deforestation destroys **habitats** and reduces **biodiversity**: species become **extinct** and survivors show reduced **variation**. Humans may have had uses for some lost organisms.

The carbon cycle and the demand for wood

D–C

- All living things **respire** and give out carbon dioxide. In photosynthesis, plants take in carbon dioxide, but deforestation 'unlocks' it.

- In 20 years, half the tropical rainforests of Borneo have been destroyed – an area larger than Great Britain – to grow palms for oil to make soap, cosmetics and other products, and to export timber.

The carbon cycle.

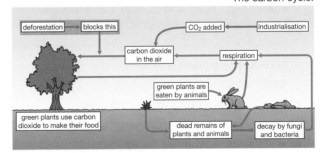

- This raises CO_2 levels in the air and threatens over 40 species of mammals, including the orang-utan.

Questions

1 Name a group of organisms that can be used as indicators of sulfur dioxide pollution.

2 If a river contains rat-tailed maggots, is it polluted or unpolluted?

3 State how deforestation affects carbon dioxide levels in the atmosphere.

4 How can rotting trees affect carbon dioxide levels in the atmosphere?

The greenhouse effect – good or bad?

The greenhouse effect and global warming

- The **atmosphere** ensures that not all the Sun's heat is reflected back into space. It traps (absorbs) some heat in the **greenhouse effect**.

- **Global warming** occurs when the greenhouse effect *increases* and less heat escapes into space. Flooding, storms and droughts will increase.

- Global warming increases with an increase in the level of the heat-absorbing gases: **carbon dioxide** from burning fossil fuels, and **methane** from cattle and rice fields.

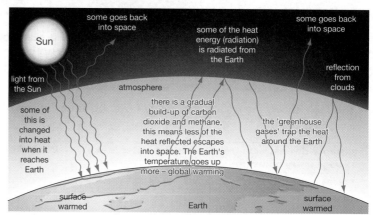

How an increase in the greenhouse effect leads to global warming.

- Air from the past trapped in Antarctic ice tells us that 250 years ago the atmosphere contained 270 parts per million of carbon dioxide. This is expected to double by 2025.

- The Earth's average temperature has risen in step with this increase. It is expected that:
  - climates throughout the Earth will change
  - Earth's great ice caps, including those at the poles, will melt, raising sea levels and flooding coasts.

Top Tip!

The greenhouse effect is good because without it the Earth would be too cold for life. But if it increases, that is *not* good.

G–E

D–C

Sustainability – the way forward?

Means of sustainable development

- As populations around the world increase and expect a better life, the Earth's **natural resources**, including fossil fuels, are being used up faster than they can be replaced or substituted.

- To ensure the long-term survival of humans, ways of **sustainable development** could include:
  - **recycling** rubbish and making it a **resource**, not waste
  - **insulating** buildings to reduce the heat energy lost
  - **walking** and **cycling** to reduce the use of cars that burn fuel
  - using **public transport** where possible.

- Fossil fuels used to generate electricity could be replaced by energy sources that do not produce carbon dioxide, such as nuclear power, wind and wave power and hydroelectric power.

How Science Works

You should be able to:
- evaluate methods used to collect environmental data and consider their validity and reliability as evidence for environmental change
- weigh evidence and form balanced judgements about some of the major environmental issues facing society, including the importance of sustainable development.

G–E

D–C

Questions

(Grades G-E)
1 Name **two** gases that cause the greenhouse effect.

(Grades D-C)
2 What has happened to CO_2 levels and temperature in the past 250 years?

(Grades G-E)
3 Explain what is meant by 'sustainable development'.

(Grades D-C)
4 State **two** alternative energy sources for generating electricity that do no produce CO_2.

B1b summary

Animals and plants are adapted:
- to **survive** in different **habitats**
- to **compete** for resources
- for **protection** against predators.

Animal and plant adaptations

Plants compete for light, water and nutrients from the soil.

Caribou and camels have adapted to extreme environments.

Animals compete for food, territory and a mate.

Reproduction can be:
- **sexual** (using **gametes**, variation in offspring)
- or **asexual** (one parent, identical offspring).

DNA contains **genes** on **chromosomes** that control our **inherited characteristics**.

Reproduction and genes

Genetically identical **clones** can be produced:
- from **cuttings** and **tissue culture** in plants
- by **embryo transplantation** in animals.

In **genetic engineering**, a gene from one organism is inserted into another organism.

Genetically modified organisms include soya plants resistant to herbicide and bacteria that make human insulin.

Charles Darwin proposed the theory of **evolution** of **species** by **natural selection**.

Mutation of a gene to cause an advantageous change to a characteristic can increase a species' chance of **survival**.

Evolution

Comparing **fossils** with present-day plants and animals can show us how species have evolved.

Factors that contribute to **extinction** include:
- environmental change
- new predators or diseases
- successful competitors
- adverse human impact on habitats.

Increase in the human **population** has affected the environment by:
- using up resources more rapidly
- increasing waste and pollution.

Land is used up for:
- buildings and quarrying materials to make them
- producing food
- waste disposal.

Intensive farming methods and **deforestation** destroy habitats, reducing **biodiversity**.

Water pollution is caused by toxic chemicals, raw sewage and excess fertiliser.

Human impact on the environment

Air pollution caused by burning fossil fuels results in smoke, smog and acid rain.

Sulfur dioxide and **oxides of nitrogen** dissolve in rainwater to give **acid rain** which damages plants, pollutes lakes and erodes limestone and sandstone structures.

Lichens and invertebrates can be used as **pollution indicators**.

Increases in the heat-absorbing gases responsible for the **greenhouse effect** can contribute to **global warming**.

Sustainable development includes using renewable energy sources, recycling and conserving energy and fuels.

Elements and the periodic table

The elements and patterns in properties

- All substances are made of tiny **particles** called **atoms**. An **element** is made of only one sort of atom.

- Atoms of each element are given a **chemical symbol**. O represents an oxygen atom and Na represents a sodium atom.

- Elements are arranged in the **periodic table** below. Each vertical column is a **group**. Elements in a group have similar properties.

- In 1869, Mendeleev first made a similar table of the elements known then. He left gaps for unknown elements, which scientists later discovered.

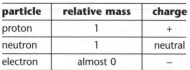

The periodic table of the elements.

Atomic structure 1

Particles in an atom

- Each atom has a small central **nucleus**. It contains **protons** and **neutrons** (except hydrogen; its nucleus is just one proton).

- **Electrons** orbit around the nucleus.

- Protons, neutrons and electrons are unimaginably small and impossible to weigh. But we use numbers, **atomic mass units**, to express their masses relative to one another (**relative masses**).

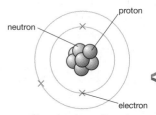

The structure of an atom.

particle	relative mass	charge
proton	1	+
neutron	1	neutral
electron	almost 0	−

- In an atom, the number of protons = number of electrons = **atomic number** (unique to each element). An atom has no net charge.

- The **mass number** = number of protons + number of neutrons

- The electrons are in **shells**. Two electrons fill shell 1 and eight electrons fill shell 2.

Questions

(Grades G-E)
1 How many different kinds of atoms are there in an element, such as oxygen?

(Grades D-C)
2 Use the periodic table to find the chemical symbol for zinc.

(Grades G-E)
3 Name the **two** kinds of particles found in the nucleus of an atom.

(Grades D-C)
4 Use the periodic table to find the atomic number of magnesium. What does this tell you about the structure of a magnesium atom?

Bonding

Bonds in elements and compounds

- Few elements exist naturally as single atoms. Two atoms of the *same* element can join to form a **molecule**:
  - nitrogen and oxygen molecules, N_2 and O_2, each have two atoms
  - the atoms *share* a pair of **electrons**, one from each atom, in a **covalent bond**.

- When atoms of two (or more) *different* elements join, they form a **compound**.
  - The atoms in water, H_2O, are arranged as HOH. Each hydrogen atom *shares* a pair of electrons with the oxygen atom. Water is a covalent compound.
  - A magnesium atom and an oxygen atom join to form magnesium oxide, MgO. Magnesium *gives two* electrons, and oxygen *takes* them. Both magnesium and oxygen then have a charge – they are **ions**. Their opposite charges attract, forming an **ionic bond**.

- Atoms with a full outer shell of electrons (see page 26) are **stable**. The atoms of most elements have an incomplete outer shell.

- An atom will react chemically with another atom that either gives, takes or shares electrons to give it a complete outer shell.

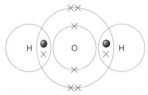

All atoms in a water molecule have full outer electron shells: the hydrogens have two electrons in shell 1 and oxygen has eight electrons in shell 2.

Extraction of limestone

Limestone: its source and uses

- **Limestone** is the chemical compound **calcium carbonate**. It is the main compound in chalk and marble, which, like limestone, are mined from quarries.

- Limestone is a building material. It is also used in the making of cement, concrete and glass and in extracting iron from its ore.

How Science Works

You should be able to: consider and evaluate the environmental, social and economic effects of exploiting limestone and producing building materials from it.

Limestone extraction: for and against

- Advantages of a local **limestone quarry** include:
  - jobs for local people
  - a better road system
  - local income to pay for health care, recreational and other facilities.

- Disadvantages include:
  - scarred landscape and loss of natural habitats
  - heavy traffic; lorries shake and damage buildings and roads
  - noise from blasting and dust pollution.

- Remedies to minimise the disadvantages include: use of rail transport, daytime working on weekdays only, and landscaping spent quarries.

Quarries can cause environmental damage.

Questions

1 Nitrogen gas in the air exists as pairs of nitrogen atoms joined together. Is nitrogen gas an element or a compound?

2 An atom of oxygen has six electrons in its outer shell. How many does it need to make this shell full?

3 Name the chemical compound in limestone.

4 List **two** advantages and **two** disadvantages of having a limestone quarry near a small town.

Thermal decomposition of limestone

Reaction of limestone on heating

- The **formula** of a compound shows the number and type of atoms joined together.

- Limestone is mainly **calcium carbonate**, formula **CaCO₃**. It contains 1 calcium atom, 1 carbon atom and 3 oxygen atoms.

- When limestone is heated strongly, a **thermal decomposition** reaction occurs – a breakdown into simpler compounds:

  calcium carbonate → calcium oxide (quicklime) + carbon dioxide

- On heating, other metal carbonates break down similarly:
  magnesium carbonate → magnesium oxide + carbon dioxide

- Calcium oxide (quicklime) reacts with water to produce calcium hydroxide (slaked lime):

  calcium oxide + water → calcium hydroxide

G–

The equation for the thermal decomposition of limestone

- The **symbol equation** for the chemical reaction of limestone on heating is: $CaCO_3 \rightarrow CaO + CO_2$

- The equation is **balanced**, having the same number of each type of atom on either side.

- In this and all reactions: no atoms are lost or made; the mass of the products equals the mass of the reactants (in this case, one reactant).

D–C

Uses of limestone

How Science Works

You should be able to:
- evaluate the developments in using limestone, cement, concrete and glass as building materials
- evaluate their advantages and disadvantages over other materials.

Limestone in building materials

- **Cement** is formed when crushed **limestone** is strongly heated with **clay** in a **kiln**. Cement reacts with water and sets gradually, and eventually becomes very hard.

- To make **concrete**, sand and rock chippings are added to cement. The mixture reacts with water and sets hard.

- Bricks are joined together with **mortar**, a mixture of sand, cement and water that sets hard.

- **Slaked lime** (**calcium hydroxide**), once widely used for building, has been mostly replaced by cement which lasts longer.

- To make **glass**, **sand** and **sodium carbonate** are heated with limestone.

G–E

Mortar and concrete

- When mixed with water, the cement in **mortar** 'hydrates', hardening and binding the sand particles together.

- **Concrete** is a mixture of cement, sand and rock chippings, materials of different sizes. When cement sets, these materials together have a far greater strength than mortar.

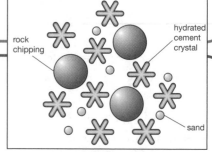

In concrete, hydrated cement crystals bind together the rock chippings and sand.

D–C

Questions

Grades G-E
1 Write down the word equation for the thermal decomposition of limestone.

Grades D-C
2 Write the balanced equation for the thermal decomposition of limestone.

Grades G-E
3 How is glass made?

Grades D-C
4 What is the difference between concrete and cement?

The blast furnace

Raw materials for making iron

- **Iron** is a strong, very widely used metal, often in the form of **steel**. It is used in cars, trains and ships, and in countless other products (see page 30).

- The raw materials for making iron are abundant and economical to extract and use:
  - **haematite**, iron ore, mainly **iron(III) oxide**, Fe_2O_3
  - **coke**, purified carbon-rich coal
  - **limestone**, mainly calcium carbonate, $CaCO_3$
  - **oxygen** in air.

Haematite contains the compound iron(III) oxide (Fe_2O_3).

Extracting iron from its ore

- Iron is extracted from its ore in a blast furnace.

- Iron is extracted by heating it with carbon, because carbon is more reactive than iron.

- The carbon takes oxygen from the iron ore, so **reducing** it.

- Iron ore, coke (carbon) and limestone go into the top of the blast furnace, and hot air into the bottom.

- The limestone helps to remove impurities from the iron ore.

- **Molten iron** is produced in the blast furnace.

Using iron

Different types of iron

- Iron from the blast furnace is 96% iron. It is remelted and moulded to form **cast iron** objects which require **hardness** and **strength** (e.g. drain covers).

- **Impurities** that make cast iron **brittle** and limit its use are removed to give **wrought iron**. This is pure iron which is softer and easily bent and shaped.

- Most iron produced is purified and then made into types of **steel**. **Carbon** and other **metals** are added in different quantities, according to the properties the steel needs to have.

The structure of wrought iron

- **Wrought iron** is pure iron metal. The atoms are bonded together in a three-dimensional, **regular**, closely packed arrangement called a **lattice**.

- Wrought iron is relatively soft and easily shaped because **layers** of atoms can slide over each other. This makes it suitable for gates and railings but too soft for many other uses.

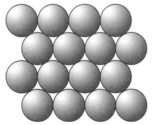

The regular arrangement of atoms in wrought iron.

Top Tip!

Pure metals are soft and easily shaped because the atoms form a regular arrangement. The layers of atoms can then pass easily over each other.

Questions

(Grades G-E)
1 What is the name of the iron compound found in iron ore (haematite)?

(Grades D-C)
2 Iron ore is *reduced* in the blast furnace. What does this mean, and why does it happen?

(Grades G-E)
3 What is done to the iron from a blast furnace, to make it into wrought iron?

(Grades D-C)
4 Explain why pure metals, such as iron, can easily be shaped.

Using steel

Types of steel

- Steel is an **alloy** of iron. Different types of steel contain other metals and carbon (a non-metal) in different amounts.

- Compared with pure iron, steel is harder, much stronger, resists corrosion and is not brittle.

The hammer must be strong (not brittle). The scissors must be hard to retain its cutting edge, and be rust-proof.

Low, medium and high carbon steels

- The different elements in alloys have different sized atoms that distort the layers of atoms in the pure metal. They cannot slide easily over each other, so alloys are **hard**.

- *Low* carbon steels (0.4% carbon or less) are soft and easy to shape, properties suitable for car body making.

- *Medium* carbon steels (about 0.8% carbon) are harder and stronger, suitable for hand tools.

- *High* carbon steels (1.0–1.5% carbon) are very hard, suitable for knives and razor blades, but may snap when bent.

- *Stainless* steel is 70% iron, 20% chromium and 10% nickel. It is very resistant to corrosion (rust).

Transition metals

Remember that metals are on the left-hand side of the periodic table, non-metals are on the right.

The properties of transition metals

- In the periodic table (see page 26), the **transition metals** are all the elements in the nine columns headed scandium (Sc) to copper (Cu).

- Transition metals are strong and make excellent **structural** materials.

- They are good **conductors** of **heat** and **electricity**, and can be **drawn** into wires and **hammered** into sheets.

How properties of metals relate to their structure

- The outer shells of transition elements contain **delocalised** electrons *free* to move through the metal lattice (see page 29).

- They leave the metal atoms with a positive charge, so they are actually ions.

- Delocalised electrons and ions attract each other, holding the structure together.

- The ordered layers of atoms in pure metals enable them to be **drawn** into wires and hammered into shapes.

- Metals conduct **heat** well. The mobile delocalised electrons carry thermal energy rapidly through the metal, and transfer it to the ions.

- Delocalised electrons make metals good conductors of **electricity** because they carry charge and are mobile.

Copper can be drawn into wires.

Questions

(Grades G-E)
1 What is mixed with iron, to make steel?

(Grades D-C)
2 Explain why steel containing a lot of carbon is used for making knives.

(Grades G-E)
3 Use the periodic table to find out whether these are transition metals or not: calcium; cobalt; rubidium.

(Grades D-C)
4 Explain why copper is a very good conductor of electricity.

Aluminium

Aluminium extraction and uses

- Aluminium ore, **bauxite** (aluminium oxide), is very abundant.

- Extraction involves **electrolysis**, a process with high energy costs.

- Pure aluminium is relatively soft. Adding small amounts of similar metals makes **alloys** that are hard, strong, light (low density), easy to shape, good thermal and electrical conductors and resist corrosion.

- Aluminium alloys are used in drink cans, bicycles, cars, pylons and cooking utensils.

Electrolysis, properties and atomic structure

- Aluminium (unlike iron) is more reactive than carbon, so carbon will not reduce aluminium oxide.

- For electrolysis, reactants are molton to allow ions to move. Adding cryolite reduces bauxite's melting point.

- **Cathode**: **aluminium ions + electrons → aluminium atoms**. Molten aluminium sinks and flows out.

- **Anode**: **oxide ions – electrons → oxygen molecules**. Oxygen reacts with the graphite (carbon) anodes which are replaced regularly.

- Aluminium atoms pack less closely than iron atoms, so aluminium is less **dense**.

- Alloys are **hard** because different-sized atoms stop layers of atoms easily sliding over each other.

- A thin film of **aluminium oxide** quickly forms on the surface. It is a barrier to oxygen reaching deeper.

> **Top Tip!**
>
> Remember 'oil rig':
> **O**xidation **I**s **L**oss
> **R**eduction **I**s **G**ain
> (of electrons).

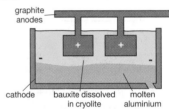

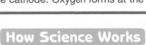

graphite anodes

cathode   bauxite dissolved in cryolite   molten aluminium

Aluminium is extracted by electrolysis. Aluminium forms at the cathode. Oxygen forms at the anode.

Aluminium recycling

> **How Science Works**
>
> You should be able to: consider and evaluate the social, economic and environmental impacts of exploiting metal ores, of using metals and of recycling metals.

Advantages of recycling

- Recycling aluminium:
  - saves the high **energy** cost of electrolysis (recycling costs are one-twentieth of this)
  - reduces the demand for **bauxite** from mining that damages the **environment**
  - reduces **waste** buried in **landfill** sites
  - can be repeated more than once.

The impact of bauxite mining

- Bauxite is mined from huge open-cast **pits**. Large areas of **trees** are felled and burned, producing carbon dioxide that contributes to **global warming**.

- In some countries, indigenous people are **moved** from areas that are their livelihood. From mine workers they contract **diseases** to which they have no immunity. The area becomes a dump for **pollutants** such as oil. There are also major dust problems in some bauxite plants in the Caribbean – people say they get asthma attacks.

Questions

(Grades G-E)
1 Name the ore from which aluminium is obtained.

(Grades D-C)
2 Why cannot aluminium be extracted from its ore by heating with carbon, as iron can?

(Grades G-E)
3 List **two** advantages of recycling aluminium.

(Grades D-C)
4 List **two** ways in which bauxite mining can harm the environment.

Titanium

Properties and extraction of titanium

- Titanium is light, strong, withstands high temperatures, resists corrosion and is easily shaped.

- Because of high energy costs for extraction of titanium from its ore, titanium dioxide, its use is limited to high-value items including replacement joints, aircraft and missiles.

Why is the F22 made from a titanium alloy?

- Titanium is more **reactive** than carbon, so carbon cannot reduce titanium dioxide (TiO_2). Because titanium dioxide is covalently bonded, so does not conduct electricity, titanium cannot be extracted by electrolysis.

- Instead, titanium dioxide is converted to titanium chloride. Then, the more reactive metal magnesium **displaces** titanium from titanium chloride.

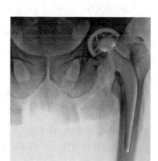

ore heated with carbon and chlorine	titanium chloride reacts with molten magnesium

$$TiO_2 \longrightarrow TiCl_4 \longrightarrow Ti$$

titanium dioxide   titanium chloride   titanium

- Titanium is very reactive but does not corrode because a thin layer of titanium dioxide forms on its surface, protecting the metal beneath.

Titanium alloy is used in a replacement hip to make the ball and shaft that fits into the top of the femur (leg bone).

Copper

Properties and extraction of copper

- Copper is a dense, **unreactive** metal with a high melting point. It is a good conductor of heat and a very good **conductor of electricity**.

- Copper wire is used for electrical wiring and plumbing.

- Ores include **chalcocite** (Cu_2S) and **chalcopyrite** ($CuFeS_2$).

- There are only limited amounts of copper-rich ores.

- The ores, extracted from open-cast pits, are low in copper, so large quantities of ore produce little copper but cause extensive damage to the landscape.

- In a traditional extraction process, the ore is crushed, washed and heated in air.

- Drawbacks are that it produces copper that must be purified, and sulfur dioxide which pollutes the atmosphere.

- In a method to increase the yield of copper from low-grade ore, the copper is first **leached** out of the ore – chemically treated to form a solution containing copper. The solution then undergoes electrolysis, and copper collects at the cathode.

Questions

(Grades G-E)

1 How do the properties of titanium make it especially useful for making aircraft?

(Grades D-C)

2 Explain why titanium cannot be extracted from its ore (titanium dioxide) by electrolysis.

(Grades G-E)

3 Why is copper used for making electrical wires?

(Grades D-C)

4 Explain why people are researching new ways of extracting copper from low-grade ores.

Smart alloys

Summary: metals and alloys

- Most **metals** are reactive, existing naturally as **compounds** in ores from which the pure metals are extracted. Gold is unreactive, occurring naturally as an **element**.

- Pure metals are usually too soft for use. Adding other metals gives **alloys** with improved properties. Steels are iron alloys (page 30) with carbon and often other metals. For aluminium alloys see page 31.

How Science Works

You should be able to:
- evaluate the benefits, drawbacks and risks of using metals as structural materials and as smart materials
- explain how the properties of alloys (other than smart alloys) are related to models of their structures.

- **Smart alloys** have a **shape memory**. They can be **deformed**, but heating returns them to their original shape.

- Two types are: the cheaper copper-zinc-aluminium alloys, and the more expensive nickel-titanium alloys used in spectacle frames.

- General properties of **transitional metals**: they are strong, can be drawn into wires and hammered into shape, and are good conductors of heat and of electricity.

metal (cost)	ore	extraction	properties and uses
iron (cheap to extract)	haematite, iron(III) oxide, Fe_2O_3	Reduction of ore with carbon monoxide: carbon monoxide + iron oxide → carbon dioxide + iron	Wrought iron: easily shaped. Steels are rigid, very strong and hard, so used for construction, bodywork, tools.
aluminium (costly to extract)	bauxite, aluminium oxide, Al_2O_3	By electrolysis of molten ore. Aluminium is more reactive than carbon, so cannot be extracted by heating with carbon (coal).	Light, does not corrode, so used for aircraft, spacecraft, drinks cans, bike frames.
titanium (very costly)	rutile, titanium dioxide, TiO_2	Many-stage process. Titanium cannot be extracted by heating ore with carbon (coal) or by electrolysis (since covalently bonded).	Very strong and light, does not corrode, so used for replacement joints, rockets, missiles and aircraft.
copper (moderately costly)	chalcocite, copper sulfide, Cu_2S	Electrolysis of solutions containing copper compounds.	Ductile (can be drawn into wires), very good conductor of electricity. Used for electrical wiring and water piping.

Fuels of the future

How Science Works

You should be able to: evaluate developments in the production and uses of better fuels, for example ethanol and hydrogen.

Fuel use: impact and remedies

- Oil is a **non-renewable** fuel source. Its supply is **finite** – it will run out eventually.

- **Petrol** and **diesel** burn to give **carbon dioxide** and **water vapour**.

- Incomplete combustion releases **unburnt hydrocarbons**, **carbon monoxide** and **nitrogen oxides**. Each causes problems for health and the environment (see also page 35).

- **Air pollution** by vehicle exhaust gases and particles is worst in cities with traffic congestion.

- New cars have a **catalytic converter** to reduce harmful emissions by converting: carbon monoxide to carbon dioxide, unburnt hydrocarbons to carbon dioxide and water, and nitrogen oxides to nitrogen. A catalytic converter works only when hot and only with unleaded petrol.

- **Ethanol** (**alcohol**), a **renewable** resource, produces less carbon monoxide than petrol. Brazil adds ethanol made from sugar cane to petrol.

- Burning **hydrogen** as a vehicle fuel causes no pollution, producing water vapour only. But hydrogen is produced by electrolysis (see page 31), a costly process.

Questions

(Grades G-E)

1 Explain why gold can be found naturally as an element, not a compound.

(Grades D-C)

2 Explain what is meant by a 'smart alloy' and give an example.

(Grades G-E)

3 Why is oil said to be a *non-renewable* fuel source?

(Grades D-C)

4 Why would using hydrogen as a fuel cause less air pollution?

Crude oil

Crude oil and fractional distillation

- **Crude oil** is a **mixture** of hundreds of compounds. Most are **hydrocarbons** whose molecules contain **hydrogen** and **carbon** atoms only.

- The size of the molecules influences the **properties** of hydrocarbons and how they are used.

- Crude oil **fractions** contain molecules with a *similar* number of carbon atoms (see table).

fraction	no. of carbon atoms	uses
petroleum gas	1–4	heating and cooking
naphtha	5–9	making other chemicals
petrol	5–10	motor fuel
kerosene	10–16	jet fuel
diesel	14–20	diesel fuel and heating oil
oil	20–50	motor oil
bitumen	over 50	bitumen (tar)

- The hydrocarbons in a fraction have similar boiling points, and each fraction has its own range of boiling points. **Fractional distillation** uses the physical process of boiling crude oil to separate the fractions.

- Thermal energy (heat) weakens intermolecular forces of attraction.

- Forces of attraction between smaller hydrocarbon molecules are less than forces between larger hydrocarbon molecules.

- So smaller molecules boil at lower temperatures than larger molecules.

- Crude oil is heated to **evaporate** all its compounds. The vapour rises up the **fractional distillation column**. As it cools, the fractions **condense** at different temperatures and leave the column.

A fractional distillation column. Temperatures are highest at the foot of the column. The smallest molecules reach the top.

Alkanes

The atomic model of methane.

The alkane family

- The **alkane** family of hydrocarbons has a **general formula** of C_nH_{2n+2}. n is the number of carbon atoms.

- The first two alkanes are: **methane** CH_4; **ethane** C_2H_6.

- Like all hydrocarbons, the atoms in alkane compounds are held together by strong **covalent** bonds (see page 27).

- Unlike some other hydrocarbon families, all the bonds in alkanes are **single** bonds. Therefore, alkanes are **saturated** compounds.

- All alkanes share a similar structure and have similar properties.

Top Tip!

Remember that all members of the alkane family have the formula C_nH_{2n+2}.

name	formula	structure
methane	CH_4	H–C–H
ethane	C_2H_6	H–C–C–H
propane	C_3H_8	H–C–C–C–H
butane	C_4H_{10}	H–C–C–C–C–H
pentane	C_5H_{12}	H–C–C–C–C–C–H

Questions

(Grades G-E)

1 Name the **two** elements that are present in hydrocarbons.

(Grades D-C)

2 Which hydrocarbon fractions come out at the top of the fractionating column – those with big molecules, or those with small molecules?

(Grades G-E)

3 Write down the chemical formula for ethane.

(Grades D-C)

4 Write down the chemical formula for an alkane with **three** carbon atoms.

Pollution problems

The consequences of burning fossil fuels

G–E

- **Coal** (carbon), **oil** and **natural gas** (hydrocarbons) are fossil fuels. When burnt in air:

  carbon + oxygen → carbon dioxide
  hydrocarbon + oxygen → carbon dioxide + water vapour

  – Carbon dioxide is a **greenhouse gas** (see page 24). In excess, it contributes to **global warming**.

- All fossil fuels contain small amounts of sulfur. When burnt:

  sulfur + oxygen → sulfur dioxide

  – Sulfur dioxide causes acid rain (see page 22).

- Tiny particles from burnt fuels may cause **global dimming**.

Acid rain, global warming and global dimming

D–C

- Normal rainwater is slightly acidic (pH 5.5):

  carbon dioxide + water → carbonic acid

- Sulfur dioxide makes rainwater more acidic (to as low as pH 2.5):

  sulfur dioxide + water → sulfuric(IV) acid (= sulfurous acid)

  – As **acid rain**, it damages vegetation and pollutes lakes and rivers, killing fish and other organisms.

- In excess, carbon dioxide unbalances the **greenhouse effect** which sustains life on Earth (see page 24), and contributes to **global warming**.

- Atmospheric smoke particles reduce the amount of sunlight reaching the Earth's surface (**global dimming**), possibly lowering its temperature.

Top Tip!
Remember that pH *decreases* as rainwater becomes acidic.

Reducing sulfur problems

How Science Works
You should be able to:
- evaluate the impact on the environment of burning hydrocarbon fuels
- consider and evaluate the social, economic and environmental impacts of the use of fuels.

Capturing the sulfur in fossil fuels

G–E

- **Sulfur** is a valuable chemical **resource**. When fossil fuels are burned, sulfur is lost to the atmosphere as sulfur dioxide.

- The sulfur dioxide formed when **coal** burns can be collected from the waste gases.

- Sulfur compounds in **natural gas** and **oil** can be removed before the fuels are burned.

D–C

- Sulfur occurs as **hydrogen sulfide** gas. A **solvent** is used to dissolve it out of natural gas and oil.

Coal-burning power stations harvest sulfur

D–C

- Coal burned in power stations produces sulfur dioxide in its waste gases:

  sulfur + oxygen → sulfur dioxide

- In **scrubbers**, the waste gases are sprayed with a mixture of powdered limestone and water. About 95% of the sulfur dioxide reacts to form solid calcium sulfate. This is collected and can be sold as a soil improver.

Sulfur dioxide is removed from the waste gases of power stations.

Questions

(Grades G-E)
1 Coal contains more sulfur than natural gas. Explain why this makes burning coal more likely to cause pollution than burning natural gas.

(Grades D-C)
2 Why is normal rain, formed in unpolluted air, slightly acidic?

(Grades G-E)
3 Give **two** reasons for removing sulfur from waste gases formed by combustion of fossil fuels.

(Grades D-C)
4 Describe how sulfur is removed from waste gases at coal-burning power stations.

C1a summary

All substances are made from **atoms**, which have:

- **protons** and **neutrons** in a small **nucleus**
- **electrons** orbiting round the nucleus.

An **element** consists of one type of atom. Two atoms of the same element can join together to form a **molecule**.

A **compound** consists of two or more different elements joined together.

Atoms and rocks

A **covalent bond** is formed when atoms **share** electrons.

An **ionic bond** is formed when atoms lose or gain electrons to become oppositely charged **ions** that attract each other.

Atoms with a full outer **shell** of electrons are **stable**.

Limestone is **calcium carbonate**.

When heated, it undergoes **thermal decomposition**:

calcium carbonate → calcium oxide
                (quicklime) + carbon dioxide

Limestone can be used to make cement, concrete and glass.

Metals are elements that are extracted from their **ores**, often oxides, by:

- heating with carbon (iron)
- electrolysis (aluminium and copper).
- other chemical reactions (titanium).

Iron is made in a **blast furnace** from **haematite**:

- impurities make **cast iron** from the furnace hard but brittle
- removing impurities makes **wrought iron**, which is easy to shape but soft
- adding metals and carbon to pure iron makes **steel** (an **alloy** of iron).

Metals

Aluminium is extracted from **bauxite** by **electrolysis**:

- it has a low density and is soft
- it forms strong, low density alloys with other metals.

Transition metals, such as **iron**, **titanium** and **copper**:

- are found in the central block in the periodic table
- are hard and strong
- are good conductors of heat and electricity
- can be drawn into wires and hammered into sheets.

Smart alloys can be **deformed**, but return to their original shape when heated.

Crude oil is a **fossil fuel**. It is a mixture of **hydrocarbon** (i.e. hydrogen and carbon only) compounds that can be separated by **fractional distillation**.

Alkanes are an important type of hydrocarbon with the **general formula C_nH_{2n+2}**.

All bonds in alkanes are **single, covalent** bonds.

Crude oil

Burning fossil fuels releases useful **energy**, but harmful substances are also made:

- **carbon dioxide** contributes to **global warming** by increasing the **greenhouse effect**
- **sulfur dioxide** causes **acid rain**
- **smoke** particles contribute to **global dimming**.

Incomplete combustion produces unburnt hydrocarbons, carbon monoxide and nitrogen oxides. All cause problems for health and the environment.

Scientists are working to reduce pollution and develop alternative, cleaner fuels for the future, e.g. **ethanol**.

Cracking

New molecules from crude oil

G–E

- Many of the hydrocarbons in crude oil are long-chain **alkanes**. They are separated into **fractions** of similar-sized molecules.

- We describe alkanes as **saturated** because all their carbon atoms are joined by **carbon-carbon single bonds**, C–C (see page 34). This makes alkanes relatively unreactive.

- Short-chain alkanes are more useful as **fuels** – petrol and diesel (motor vehicles), fuel oil (ships) and kerosene (aircraft).

- The **cracking** process makes shorter-chain molecules from long-chain ones:
  - crude oil is heated to form a vapour that passes over a hot catalyst
  - the molecules 'crack', splitting into shorter molecules that include **alkenes**

    long-chain alkanes → shorter-chain alkanes + alkenes
  - this reaction is an example of **thermal decomposition**, in which large molecules are broken into smaller ones by heating.

> **Top Tip!**
> Monkeys Eat Peanut Butter helps you count the number of carbons in molecules.
> 1 = meth-
> 2 = eth-
> 3 = prop-
> 4 = but-

Why hydrocarbons are cracked

D–C

- The hydrocarbons required for petrol are 5 to 12 carbon atoms long. Fractions containing longer-chain molecules are cracked to make more petrol.

- Cracking a hydrocarbon molecule produces an alkane and an alkene. Alkenes are a family of more reactive hydrocarbons that are the starting materials for making many useful compounds, including **plastics**.

- Plastics are **polymers**, formed when alkene molecules join in very long chains.

Alkenes

Alkenes

G–E

- **Alkenes** are a family of hydrocarbons.

- The smallest alkene is **ethene**, chemical formula C_2H_4. The next is **propene**, C_3H_6.

- The **general formula** for the alkenes is C_nH_{2n}.

ethene   propene

Why alkenes are more reactive than alkanes

G–E

- All alkenes are **unsaturated** because they contain a **carbon-carbon double bond**.

D–C

- Both double-bonded carbons in alkenes are joined to *three* other atoms. They can react and join to *four*. So alkenes are much more **reactive** than alkanes.

- The general formula for alkenes – C_nH_{2n} – shows that for every carbon atom, an alkene has two hydrogen atoms.

> **Top Tip!**
> In all molecules, a carbon atom always forms four bonds. This is a good way to check if your structures are correct.

Questions

(Grades G-E)

1 Alkanes are saturated molecules. What does this mean?

(Grades G-E)

2 Describe how long-chain alkanes in crude oil are cracked.

(Grades G-E)

3 Alkenes are unsaturated molecules. What does this mean?

(Grades D-C)

4 Explain why alkenes are more reactive than alkanes.

Making ethanol

Making ethanol from crude oil

- Ethene is a gas produced by cracking crude oil (see page 37).

- Ethene is an **alkene**, and so it is reactive. When a mixture of ethene and steam is passed over a **catalyst**, the colourless liquid **ethanol**, C_2H_5OH, is formed.

ethanol

- Crude oil is a **non-renewable** source of ethene, and therefore of ethanol.

- Ethanol is in the **alcohol** family of compounds, containing an –OH group.

- The –OH group makes ethanol reactive. Ethanol is highly **flammable**. It burns to give carbon dioxide and water.

G–E

Making ethanol from plants

- Throughout history, sugar in grapes has been **fermented** to make ethanol ('alcohol'). The sugar is dissolved in water with **yeast**. The yeast changes the sugar to ethanol.

- Crops, including sugar cane and sugar beet, also provide sugar that is fermented to ethanol.

- In some countries including Brazil (see page 33), ethanol is added to petrol, or replaces it as a motor fuel. Ethanol is produced from **renewable** crops. It is a **biofuel**.

Top Tip!

Ethanol is the chemical name for alcohol. It has the formula C_2H_5OH. Ethanol is not a hydrocarbon because it contains oxygen as well as carbon and hydrogen.

D–C

How Science Works

You should be able to: evaluate the advantages and disadvantages of making ethanol from renewable and non-renewable sources.

Plastics from alkenes

How polymers are formed

- **Polymers** are made when small molecules called **monomers** are joined together to make very long chains.

- The polymer **poly(ethene)** (**polythene**) is made from molecules of the gas ethene formed in **cracking** (page 37). Ethene is put under very high pressure and heated with a catalyst. The reaction 'uses up' the double bond in each ethene monomer:

three ethane molecules   poly(ethene)

G–E

- The **carbon-carbon double bond** makes alkenes very reactive (see page 37). In a reaction, one of the bonds between the two carbon atoms breaks. This allows both carbon atoms to join to an additional atom.

- This is called **polymerisation**, when many thousands of monomers of one compound join up like a string of beads.

- The alkene, **propene** can polymerise to form poly(propene):

D–C

three propene monomers   a section of a strand of poly(propene)

Questions

(Grades G-E)
1 Describe how ethanol is made from ethene.
(Grades D-C)
2 Explain why ethanol is reactive.

(Grades G-E)
3 Poly(ethene) is a polymer. What is a polymer?
(Grades D-C)
4 Explain why alkenes can form polymers.

Polymers are useful

Special properties of polymers

G–E

- An artificial hip replaces the ball and socket of a worn hip joint (see page 32). The socket is made of the polymer **Teflon®**:
  - its surface has almost no friction, so the ball of titanium alloy swivels freely
  - it is non-toxic and inert (non-reactive), essential inside the body
  - it can withstand high temperatures (it is also used in non-stick pans).
- Teflon® is like poly(ethene) (polythene) but a fluorine atom replaces every hydrogen atom. The fluorine gives Teflon® its special properties.

D–C

- **Slime** is made from aqueous solutions of the polymer **poly(ethenol)** (PVA) and of the salt borax. When they mix, borax forms **cross-links** between the polymer chains.
- Slime forms a ball which seems to melt when tipped, and breaks when stretched quickly.
- The **viscosity** of slime depends on the concentrations of the solutions and the length of the polymer chains.

> **Top Tip!**
>
> Viscosity is a measure of the thickness or gooeyness of a liquid. The thicker it is, the higher its viscosity.

Uses of polymers

D–C

- Chemists choose monomers that give polymers (plastics) with the **properties** required for a particular purpose:
  - packaging materials must be strong, light and flexible
  - waterproofing for tent fabrics must repel water
  - dental polymers for crowns and dentures are designed to harden when exposed to ultraviolet light
  - wound dressings must breathe and hydrogels must retain moisture for rapid healing.

Disposing of polymers

Reasons for recycling

G–E

- Most polymers are produced from **non-renewable** crude oil. They are not **biodegradable** (microorganisms cannot break them down). They take up space in waste dumps and landfill sites.
- Polymers can be **recycled**. They are shredded, melted and formed into pellets used to make new objects such as fleeces.
- It is becoming cheaper to recycle polymers than to throw them away.

Alternatives to recycling

D–C

- Polymers can be burned to generate heat energy, but they can produce particle **pollution** and toxic fumes that damage the **environment**.
- To make them **biodegradable**, polymers are mixed with cellulose or starch that microorganisms will break down.

> **How Science Works**
>
> You should be able to:
> - evaluate the social and economic advantages and disadvantages of using products from crude oil as fuels or as raw materials for plastics and other chemicals
> - evaluate the social, economic and environmental impacts of the uses, disposal and recycling of polymers.

Questions

(Grades G-E)

1 State **two** useful properties of the polymer Teflon®.

(Grades D-C)

2 Slime is a polymer. State **one** property of it that varies according to the conditions under which it is made.

(Grades G-E)

3 Why are there problems with the disposal of waste polymers?

(Grades D-C)

4 Describe how polymers can be made biodegradable.

Oil from plants

Sources of vegetable oils

- The fruits, seeds and nuts of some plants are rich in **vegetable oils**.
- Trees with oil-rich **fruits** include palm, olive and avocado.
- The **seeds** of cotton, rape, mustard, sunflower and sesame supply oils.
- Oil-rich **nuts** include coconuts and groundnuts.
- Fruits, seeds and nuts are crushed in a press to break any hard coatings and to extract the oils. Then water and impurities are removed.

Purification and importance of vegetable oils

- To **purify** an oil containing water and other impurities, the mixture can be shaken with a hydrocarbon solvent (which does not mix with water) and distilled. Water molecules are small, while the solvent and oil molecules are larger and have higher boiling points. So, on heating, the water boils first and is discarded. Then the oil and solvent boil off together and are collected. Dust and other rubbish remain in the flask.
- Natural oils are high in chemical **energy**. They are used widely in foods and for cooking.
- Some nuts, particularly almonds, brazil nuts and pistachios, are excellent sources of **vitamins** and **trace elements**.

> **Top Tip!**
>
> Distillation works because liquids have different boiling points.

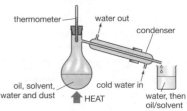

Distillation apparatus. The water boils off first, then the oil/solvent mixture.

Green energy

Biofuels and fossil fuels

- Some vegetable oils and other substances obtained from plants can be used as **fuels**, known as **biofuels**.
- Biofuels, either alone or mixed with petrol or diesel, are sometimes called **green fuels**. They:
  - include rapeseed oil, sunflower oil and wood
  - are **renewable fuels** because new plants will supply more of them
  - are 'greenhouse neutral' ('carbon neutral') because the amount of carbon dioxide *given out* when they burn is *taken in* by plants grown to replace them.
- As **non-renewable** fossil fuels run out, the need for renewable fuels will increase.

- As the supply of fossil fuels diminishes, their price rises. World economies needing alternative energy sources see biofuels as a major solution.
- To increase the use of green fuels for motor vehicles:
  - biofuels too thick for some engines are mixed with traditional fuels
  - new engines are being designed that prevent soot and the poisonous gas carbon monoxide from forming.

> **How Science Works**
>
> You should be able to:
> - evaluate the effects of using vegetable oils in foods and the impacts on diet and health
> - evaluate the benefits, drawbacks and risks of using vegetable oils to produce fuels.

Questions

(Grades G-E)
1 Give **two** sources of vegetable oil.

(Grades D-C)
2 Describe how a vegetable oil can be purified.

(Grades G-E)
3 Give **two** examples of biofuels.

(Grades D-C)
4 Explain why we will need to use more biofuels in the future.

Emulsions

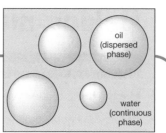

This emulsion consists of tiny oil droplets suspended in water.

Making an emulsion

- Oil and water do not mix. The oil forms a separate layer on top.
- When shaken hard, oil forms tiny droplets in water. This mixture is an **emulsion**. An emulsion is formed when one liquid is dispersed in another.
- Adding an **emulsifier** stops the oil from separating out from the water.
- Emulsions are thicker than oil or water alone, and have useful properties:
  - sauces and dressings are easy to pour slowly
  - emulsion paints cover surfaces well without dripping.
- Milk is an emulsion of fat droplets spread out in water.
- Butter is an emulsion of water droplets spread out in fat.

> **Top Tip!**
>
> Oils are liquids at room temperature, and fats are solids.

More about emulsions

- The liquids in an emulsion do not normally mix – they are **immiscible**.
- The liquid forming droplets is the **dispersed phase**. The liquid between the droplets is the **continuous phase**.
- Droplets of the dispersed phase are **suspended** in the continuous phase.
- Emulsions are opaque or milky because the tiny droplets scatter light.

> **Top Tip!**
>
> Liquids that do not mix together are called immiscible liquids. Those that do mix together are called miscible liquids.

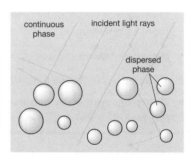

Light rays are scattered by the droplets in an emulsion, making it look opaque.

Polyunsaturates

Saturated and unsaturated oils and fats

- Unsaturated fats and oils have one or more double bonds between carbon atoms.
- Olive oil has one double bond, so it is a **monounsaturated** oil.
- Fats with more than one double bond are **polyunsaturates**.
- Fats and oils with single bonds only between carbon atoms are **saturated**.
- To find if a fat or oil is saturated or not, the **bromine water** (or iodine water) test is used.

- If there are no double bonds, the orange colour of the bromine is unchanged.
- In the presence of one or more double bonds the bromine becomes colourless. This is because bromine reacts with the carbons at the double bond.

> **Top Tip!**
>
> Do not say that the bromine becomes clear. It is clear anyway, even when it is orange! Say that it changes its colour from orange to colourless.

Questions

(Grades G-E)
1 Give **one** example of an emulsion.

(Grades D-C)
2 Explain why emulsions are not transparent.

(Grades G-E)
3 Explain the difference between a 'saturated' fat and an 'unsaturated' fat.

(Grades D-C)
4 Describe how you test a liquid to find out if it contains unsaturated compounds.

Making margarine

Making margarine

- **Margarine** is made from vegetable oils such as corn, sunflower and olive oils.
- The *liquid* vegetable oil is **unsaturated**, containing **carbon-carbon double bonds**. It is reactive and is chemically hardened to form a *soft solid* fat:
  - hydrogen gas passes through the oil warmed to about 60 °C
  - a nickel catalyst speeds up the reaction
  - hydrogen reacts with the carbon-carbon double bonds in the oil.

  The reaction is **hydrogenation**.

G–E

How hydrogenation changes vegetable oils

- The molecules of fats and oils contain long-chain molecules called **fatty acids** with carbon-carbon double bonds.
- At a double bond, fatty acid molecules bend in a V-shape. They cannot pack tightly, and the oil remains liquid at room temperature.
- **Hydrogenation** replaces some of the double bonds in the fatty acids with single bonds.
- These fatty acid molecules straighten out and can then pack closely in parallel.
- This raises the density and the melting point of the vegetable oil, and it becomes a solid fat suitable for spreading and cake-making.

Oleic acid has one double bond.

D–C

Food additives

What are additives?

- An **additive** is a chemical put in food to keep it fresh or make it appetising.
- Some additives are natural and some are synthetic. Small amounts are safe, but large amounts of some additives could be **toxic**.
- Each permitted additive has been tested for safety and is allocated an **E-number** which identifies its chemistry and what it does.
- The use of additives is regulated by the European Union and the UK government. Additives must be listed on food labels.
- Permitted types of additives include:
  - antioxidants
  - colourings
  - emulsifiers, stabilisers, gelling agents and thickeners
  - flavourings
  - preservatives
  - sweeteners.
- Permitted additives are not harmful, but none of the synthetic additives is essential for a healthy diet.

How are sweets made to look as colourful as this?

G–E

D–C

How Science Works

You should be able to: evaluate the use, benefits, drawbacks and risks of ingredients and additives in food.

Questions

Grades G-E
1 What is margarine made from?

Grades D-C
2 Explain why hydrogenation makes liquid oils become solid margarine.

Grades G-E
3 What is a food additive?

Grades G-E
4 Explain why additives listed on food labels have an E-number.

Analysing chemicals

Analysing the chemicals in food

G–E

- **Additives** in foods, such as artificial colourings, can be identified by techniques of **chemical analysis**.

- **Paper chromatography** separates substances in a mixture. Samples removed from the paper can be chemically analysed and identified. Further studies can check whether they are safe.

Separating dyes in a food dye using paper chromatography

D–C

- The preparation of a chromatogram of artificial food dye is as follows:
  - the unknown food dye is spotted on the pencil line alongside samples of known dyes
  - the bottom edge is placed in a solvent (e.g. water or ethanol) that dissolves all the dyes
  - the solvent rising up the paper carries the known dyes and separates out the food dyes
  - when the spots are well up the paper, the paper is removed from the solvent and dried
  - the food dye spots are matched to the known dyes, and unidentified spots are analysed further.

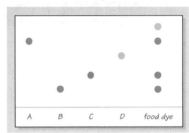

Chromatogram of known dyes and an unknown food dye.

The Earth

The Earth and its atmosphere

G–E

- The Earth is one of the rocky planets of the Solar System (see page 67) and is about 4.6 billion years old. The Moon is Earth's satellite.

- The Earth is a slightly squashed sphere with a diameter of 13 000 km surrounded by an atmosphere about 30 km deep.

- Seas and oceans cover roughly 70% of the surface. Life evolved in water some 3.4 billion years ago. Humans appeared 30 000 years ago.

land forms about 30% of the surface

diameter 13 000 km

water covers about 70% of the surface

The size and features of the Earth.

The Earth's interior

D–C

- The Earth has three main layers
  - the **crust**, the **mantle** and the **core**.
- The outermost **crust** is a thin layer of solid rock.
- The **mantle**, largely of solid rock, occupies half the diameter. Heat from radioactive decay drives convection currents that circulate the 'plastic' rock beneath the crust.
- The **core** occupies the centre and about half the Earth's diameter:
  - the *outer core* is a fluid mixture of iron and nickel; the Earth's magnetic field is thought to be generated here
  - the *inner core* is solid iron and nickel.

crust: cold, thin, solid rock layer

mantle: hot 'plastic' rock layer moved slowly by convection currents

core: hot iron and nickel, liquid outer part, solid central part

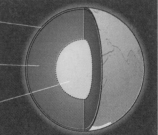

The layers of the Earth.

Questions

(Grades G-E)

1 Name a technique that can be used to separate a mixture of artificial colours in a food.

(Grades D-C)

2 How does this technique separate the colours?

(Grades G-E)

3 How old is the Earth?

(Grades D-C)

4 Name the **three** main layers that make up the Earth.

Earth's surface

Old theory of Earth's formation

- About 4.6 billion years ago, the Earth's surface solidified.
- Scientists once thought that:
  - the landmass then shrank, breaking into the pattern of **continents** and oceans that exists today
  - heat in the Earth's interior was just the heat left over from the Earth's formation.

G–E

How Earth's surface features were formed

- In 1915, the German Alfred Wegener suggested that:
  - the first landmass to form was one **supercontinent**, which he named **Pangaea**
  - Pangaea broke up 200 million years ago, and the pieces (plates) 'drifted' to their present positions.

Pangaea, over 200 million years ago.

- Knowing more about the Earth's interior, today's scientists agree with Wegener except for how he thought continental drift happens.

- The Earth's crust and the semi-solid upper mantle form the **lithosphere**. It is cracked into **tectonic plates**.

- Below the lithosphere is the 'plastic' rock of the **mantle**. Heat from radioactive processes inside the Earth drives **convection currents** in the rock. The tectonic plates ride on this circulating material.

- Movement is only centimetres a year, but, for example, the landmass of India has been carried across an ocean to collide with Asia.

D–C

Earthquakes and volcanoes

Earthquakes, volcanoes and subduction zones

- As the mantle moves underneath the plates, it exerts huge pressures on them. Eventually, plates are forced to move relative to one another where weakest – along **plate boundaries** – and an **earthquake** occurs. This usually happens very suddenly.

- A plate can push:
  - into another, driving them both upwards
  - along the boundary in a sliding action
  - under an adjacent plate which then overlaps it.

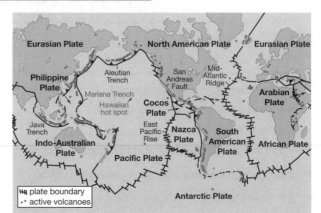
The main plate boundaries and active volcanoes.

- **Volcanoes** form at points of weakness in the Earth's crust. Hot molten **magma** from the mantle is forced up through a **vent** in the crust, and blasts out as **lava** and gases. Solidifying lava forms the volcano's cone shape round a crater.

- A region where one plate moves under another is a **subduction zone**. Subduction zones are at high risk of earthquakes and volcanic eruptions.

D–C

Questions

Grades G-E

1 What was the first explanation that scientists suggested for why the Earth has continents and oceans?

Grades D-C

2 What makes the inside of the Earth hot?

Grades D-C

3 Explain why earthquakes happen very suddenly.

Grades D-C

4 Explain why earthquakes often happen at subduction zones.

The air

Composition of the air

- The air is a **mixture** of colourless gases. It is mostly **nitrogen**, **oxygen** and a variable amount of **water vapour**.

- Other gases, in very small amounts, include **carbon dioxide** and the **noble gases**.

- The noble gases, in Group 0 of the periodic table, are unreactive because they have a full outer shell of electrons.

- They are used in filament lamps and in electrical discharge tubes.

- Helium is used in air balloons because it is much less dense than air.

- The gases in Earth's early atmosphere were very different from those of today.

- Developing life forms changed the **composition** of the atmosphere.

Gases in the air.

Evolution of the air

Gases of the present atmosphere

- The table shows the percentage by volume of the ten most abundant gases in air today.

gas	nitrogen	oxygen	argon	carbon dioxide	neon	helium	methane	krypton	hydrogen	xenon
% of air	78.08	20.95	0.93	0.035	0.0018	0.00052	0.0002	0.00011	0.00005	0.00001

Evolution of the atmosphere

- The graph shows the change during Earth's history in levels of nitrogen, oxygen and carbon dioxide.

- Intense volcanic activity of the early Earth produced **water vapour** which **condensed** to become the first seas.

- **Carbon dioxide** and **methane** also came from volcanoes.

- The very high level of carbon dioxide dropped as it dissolved in the seas.

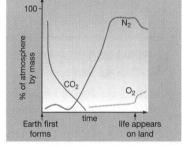

Changes in the levels of nitrogen (N_2), oxygen (O_2) and carbon dioxide (CO_2) since the Earth's formation.

- Early life forms evolved in warm shallow water which protected them from the Sun's harmful ultraviolet rays. Some of these photosynthesised. Their waste product, **oxygen**, built up in the atmosphere.

- Some oxygen was converted to **ozone**, O_3, forming a shield against ultraviolet rays. This enabled plants to evolve on land.

- Plants photosynthesised, removing carbon dioxide and adding more oxygen. Animals evolved, taking in oxygen and giving out carbon dioxide.

- **Ammonia**, NH_3, from volcanoes, reacted with oxygen gas and produced **nitrogen**, which gradually built up in the atmosphere. Bacteria also produced nitrogen from ammonia.

- The gases in the atmosphere eventually became balanced about 200 million years ago.

Questions

(Grades G-E)

1 List the **three** most abundant gases in our atmosphere.

(Grades G-E)

2 What is the approximate concentration of carbon dioxide in the atmosphere?

(Grades G-E)

3 Which of the gases shown in the table above are noble gases?

(Grades D-C)

4 Explain why there is oxygen in our atmosphere.

Atmospheric change

The atmosphere is heating up

- The atmosphere is getting warmer. This is called **global warming**, and its effect on the weather is **climate change**.

- In the past 100 years, the **average temperature** of the Earth's atmosphere has increased by 1 °C.

- As atmospheric temperature increases:
  - ice sheets melt and sea level rises, leaving low coasts and islands under water
  - the energy in the atmosphere increases and weather becomes stormier and more extreme. Generally, summers are hotter and, in some places, winters are colder.

G–E

Global warming

- Atmospheric temperature remains stable while the heat gained from the Sun is balanced by the heat lost to space (see page 24).

- Temperature rises when the level of **greenhouse gases** increases. **Carbon dioxide** is a greenhouse gas.

- Carbon dioxide circulates between the atmosphere and living and non-living things (see page 78).

- The burning of **fossil fuels** has added more carbon dioxide to the atmosphere than can leave it.

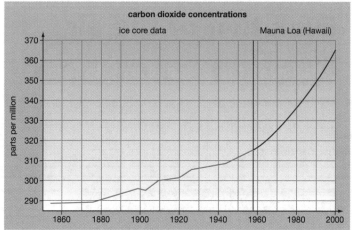

Atmospheric concentrations of carbon dioxide from 1860 to 2000.

D–C

Carbon sinks

- Carbon in carbon dioxide leaves the atmosphere when it is deposited in **carbon sinks**. These include **sedimentary rocks**, **plants** and **fossil fuels**.

- Most of the carbon in the Earth's original atmosphere is locked up in carbon sinks.

- Carbon dioxide is *water soluble*.
  - It dissolves in rainwater and seawater. It forms **carbonic acid** that reacts with calcium-containing materials in seawater to form **calcium carbonate**. Being insoluble, this builds up as sediment on the sea floor. Over millions of years, sediments become **sedimentary rocks** including **limestone** and **chalk** which can contain the shells of microscopic organisms.

- Carbon dioxide is taken up by plants in **photosynthesis** (see page 73):
  - in a given area, trees 'lock up' far more carbon than smaller plants.

- Over millions of years, plant and animal material can be fossilised:
  - plants are buried and form **coal**
  - microscopic marine animals and plants form **oil**.

D–C

How Science Works

You should be able to:
- explain and evaluate theories of the changes that have occurred and are occurring in the Earth's atmosphere
- explain and evaluate the effects of human activities on the atmosphere.

Questions

Grades G-E

1 What has happened to the average temperature of the atmosphere over the past 100 years?

Grades D-C

2 How does carbon dioxide contribute to global warming?

Grades D-C

3 What is a 'carbon sink'? Name **three** carbon sinks.

Grades D-C

4 Most of the carbon from carbon dioxide in the air is locked up in sedimentary rocks. How does this happen?

C1b summary

New substances from crude oil

Crude oil contains **hydrocarbons** including **alkanes** with the general formula C_nH_{2n+2}. Alkanes can be cracked to form a new family called the **alkenes**.

Alkenes are **unsaturated**. They contain carbon-carbon **double bonds**. The **general formula** for alkenes is C_nH_{2n}.

Polymers are made when small molecules called **monomers** are joined together to make very long **chains** in a process called **polymerisation**. Alkenes can be made into polymers.

Polymers can be used to make many **useful substances**, e.g. Teflon® and plastics for packaging materials, dental compounds, waterproof materials, etc.

Many polymers are **not biodegradable** and are produced from **non-renewable** crude oil. To dispose of polymers, they can be:
- **recycled** to make other products
- burnt to generate heat (but this can result in pollution)
- mixed with starch or cellulose to make them biodegradable.

Oil from plants

Many **plants** contain useful **natural oils** that can be **extracted**, e.g. from fruits, seeds and nuts.

Biofuels can be made from plant oils and other substances. These '**green fuels**':
- include rapeseed oil, sunflower oil and wood
- are **renewable**
- are '**greenhouse neutral**'.

An **emulsion** is formed when one liquid is **dispersed** in another.

Fats and **oils** can be **saturated** (containing only single bonds) or **unsaturated** (containing C=C double bonds). **Hydrogenation** is used to **harden** unsaturated liquid vegetable oils in the manufacture of **margarine**.

The Earth

The **Earth** has three main layers: the **crust**, the **mantle** and the inner and outer **core**.

The Earth's **lithosphere** is cracked into **tectonic plates** which ride on the plastic part of the mantle.
Earthquakes can occur at the **plate boundaries** where they collide with each other.
Volcanoes form at weak points in the Earth's crust.

The Earth's **atmosphere** has changed over millions of years. Many of the gases that make up the atmosphere came from volcanoes, e.g. water vapour, carbon dioxide and methane.
Ozone (O_3) forms a shield against ultraviolet rays.

The **air** is a mixture of mainly **nitrogen**, **oxygen**, variable amounts of **water vapour** and very small amounts of other gases such as **carbon dioxide** and the **noble gases**.

The levels of the **greenhouse gases** in the atmosphere are rising, resulting in **global warming** and **climate change**.

Sedimentary rocks, **plants** and **fossil fuels** are **carbon sinks**, locking up the carbon from carbon dioxide that was once in the atmosphere.

Heat energy

Heat and temperature

- **Heat** is **energy** that can flow from one material to another.

- **Temperature** indicates how hot something is.
  A thermometer measures temperature.

Heat energy flows from the warm surrounding air to the cold lolly.

Thermal radiation

- **Thermal (infrared) radiation** is the **transfer** of **heat energy** by rays.

- The rays are **electromagnetic waves**. They travel:
  - in straight lines
  - at the speed of light
  - through a vacuum: the Sun's radiation passes through space.

- Thermal radiation makes particles in an object vibrate.

- All materials **absorb** (take in) and **emit** (give out) thermal radiation. Hot objects emit more than cold ones.

- An **infrared camera** detects thermal radiation.

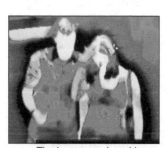

The image produced by an infrared (thermal imaging) camera shows that objects emit different amounts of thermal radiation.

Thermal radiation

Measuring emitted thermal radiation

- Hot objects always emit more thermal radiation than cold objects.

- The greater the difference between the temperatures of an object and its surroundings, the faster the rate of heat transfer.

- The amount of thermal radiation an object emits depends on its shape, dimensions and type of surface.

- A **thermopile** (e.g. an 'ear thermometer') measures thermal radiation.

- More thermal radiation is emitted by: black surfaces than white surfaces; dull surfaces than shiny surfaces.

> **Top Tip!**
> Remember that 'cold' is not transferred! It is always heat energy that is transferred, from a hotter object to a colder one.

Emission and absorption of thermal radiation

- Hot water in the dull black beaker in the diagram loses heat – **emits** thermal radiation – faster than the silver beaker.

- Filling the beakers with ice-cold water in warm surroundings shows that the dull black surface also **absorbs** heat faster than the silver surface.

- The black grid behind a refrigerator emits thermal radiation fast, while silver 'space blankets' slow down emission.

- White buildings in hot countries minimise the thermal radiation absorbed.

- As polar ice melts, the dark-coloured sea will absorb more thermal radiation and may promote **global warming**.

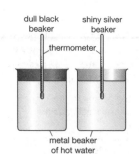

Equipment to compare the emission of thermal radiation by different surfaces.

Questions

(Grades G-E)

1 When you hold a cup of hot coffee in your cold hands, in which direction is heat energy transferred?

(Grades D-C)

2 Explain how an infrared camera can detect a person hiding among trees.

(Grades G-E)

3 Do hotter objects emit more thermal radiation or less thermal radiation than cold objects?

(Grades D-C)

4 Explain why silver objects heat up *and* cool down more slowly than black objects.

Conduction and convection

Conductors, insulators and convection

- **Heat energy** flows easily through **thermal conductors** (e.g. metals).

- It does not flow easily through **thermal insulators** (e.g. glass, plastic, air).

- A metal saucepan conducts heat rapidly from cooker to food, but a plastic handle conducts heat very slowly.

- Double-glazed windows reduce heat loss because glass layers and the air between are poor conductors.

- Warmed liquids and gases rise by **convection** because they are less dense than cooler liquids (see diagram) and gases.

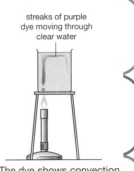

streaks of purple dye moving through clear water

The dye shows convection in the heated water.

The particle model and conduction and convection

- Particles in **solids** vibrate in fixed positions. When one end of a solid rod is heated, the particles vibrate more. They **transmit** vibrations to their neighbours, **transferring** heat energy along the rod.

- When heated, particles in **liquids** and **gases** move more vigorously. They take up more space. The heated material **expands** (becomes less dense), causing **convection currents**.

Heat transfer

Heat transfer in liquids and gases

- Liquids are poor thermal conductors – they transfer heat slowly, as the diagram shows.

- Gases are very poor thermal conductors. Air trapped in a cold-weather jacket greatly slows heat transfer from a person's body.

Top Tip!

Do not say that particles in a solid 'start to' vibrate when heated. They are already vibrating! Heat energy makes them vibrate *more vigorously*.

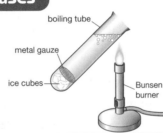

boiling tube
metal gauze
ice cubes
Bunsen burner

It takes a long time before the ice cube starts to melt. (The less dense hot water remains at the top.)

Energy transfer and the particle model

How Science Works

You should be able to evaluate ways in which:
- heat is transferred in and out of bodies
- the rates of these transfers can be reduced.

- Fixed **particles** in a solid **transfer energy** to each other by **vibrating**.

- Only particles on the surface of a solid can transfer energy to the **surroundings**. Therefore, the rate of heat transfer:
  - is fast for a small volume with a large surface area
  - is slow for a large volume with a small surface area.

- Metals are good conductors because they have **free electrons** that can move in all directions between the fixed particles. They help to transfer energy to the fixed particles.

- Particles in liquids are not firmly held together, so energy is less easily transferred from one particle to another. Heat transfer by conduction between particles is slow.

- Particles in gases are far apart. Heat energy is transferred only when particles collide, so heat transfer by conduction is very slow.

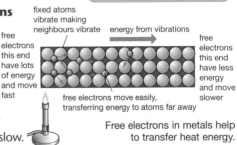

fixed atoms vibrate making neighbours vibrate

energy from vibrations

free electrons this end have lots of energy and move fast

free electrons this end have less energy and move slower

free electrons move easily, transferring energy to atoms far away

Free electrons in metals help to transfer heat energy.

Questions

Grades G-E

1 Why does hot air move upwards?

Grades D-C

2 Explain how heat energy is conducted along a metal rod.

Grades G-E

3 Why does wearing several layers of thin clothes keep you warmer than wearing one thick layer?

Grades D-C

4 Explain why gases are very poor conductors of heat.

Types of energy

Energy, forms and measurement

How many types of energy have you used today?

- **Energy** enables things to happen. Without energy, nothing happens.

- We experience energy in different **forms** (types), as **heat**, **light** or **sound**, **chemical** energy, **microwave** and **radio wave** energy, and **kinetic** and **potential** energy.

- Energy is measured in **joules (J)**. Lifting a small apple from floor to table takes about 1 J. Walking 1 km on level ground uses about 280 000 J.

- A teenager uses about 10 million J of energy a day, mostly for life processes (e.g. maintaining body heat).

- Energy can be stored for use when required. **Energy stores** include fuels, batteries, stretched elastic bands and tightly wound springs.

Energy changes

Energy transfers and transformations

- **Energy transfers** occur when the *same* form of energy moves from *one place to another*.
  For example:
  - a cyclist's **kinetic energy** is transferred to become the bicycle's **kinetic energy**.

- **Energy transformations** occur when *one* form of energy *changes to another*.
  For example:
  - a light bulb transforms **electrical** energy to **light** energy
  - a music system transforms **electrical** energy to **sound** energy
  - a battery transforms **chemical** energy to **electrical** energy.

Examples of energy transformations.

Wasted energy and the Law of Conservation of Energy

- Whenever a device transfers or transforms energy for a useful purpose, some energy is always **wasted** – it is in a form that is not useful:
  - a light bulb also produces unwanted heat
  - a rocket motor that provides kinetic energy also produces unwanted heat, light and sound.

- Energy never disappears. The total amount of energy before and after an event is always the same. The **Law of Conservation of Energy** says:

  Energy cannot be created or destroyed.

  It can only be transferred or transformed from one form to another.

  The total energy always remains constant.

Questions

(Grades G-E)

1 What form of energy do the people have because of their position on the high wall?

(Grades D-C)

2 If you use 10 million J of energy in a day, how much chemical energy should there be in the food you eat each day?

(Grades G-E)

3 A girl hits a gong with a mallet. Kinetic energy in her arm becomes kinetic energy in the gong and then sound energy in the air. Which is an energy transfer? Which is an energy transformation?

(Grades D-C)

4 When you ride a bike, not all the kinetic energy in your legs is transferred to kinetic energy in the bike. Where does the wasted energy go?

Energy diagrams

Diagrams showing energy changes

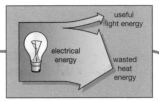

Energy changes for a light bulb.

- A **Sankey diagram** shows the **energy input** and **energy output** of a device.

- The energy output includes **useful energy** and **wasted energy**. Together, they add up to the value of the energy input. The total amount of energy stays the same.

- The second diagram shows this for an engine. Input and output are both 1000 J.

- You can use Sankey diagrams to calculate missing values for energy input and output.

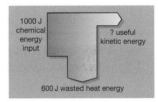

A Sankey diagram for an engine.

Measuring wasted energy

Energy transformations in different light sources.

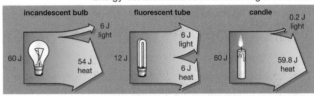

- The top diagram shows that the candle wastes the most energy and the fluorescent light wastes the least.

- A device may have several **energy transfers** or **transformations**. At each one, energy is wasted. The bottom diagram shows this for a petrol engine.

A Sankey diagram for a petrol engine.

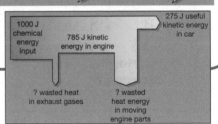

Energy and heat

Wasted heat energy and where it goes

- In any energy transfer or transformation, wasted energy eventually appears as **heat energy**.

- Wasted heat energy is not 'lost' but is transferred to the **surroundings**, making them warmer. For example, a runner who transforms **chemical energy** in glucose to **kinetic energy** for muscle contraction, will also transfer heat to the surrounding air.

- Wasted heat is transferred to the surroundings by conduction, convection and thermal radiation.

The particle model: warming and energy spread

- Inside all objects, the particles are **vibrating**. When an object absorbs any sort of energy, its particles vibrate more vigorously – it heats up. The more energy it absorbs, the more vigorously its particles vibrate and the hotter it gets.

How could you decide which is the more concentrated store of energy – the coal or the wood?

- In all energy transfers and transformations, energy spreads out and becomes more difficult to use for further energy transformations.

- Energy is more useful when **concentrated**, as in fuels and batteries, which are useful **stores of energy**.

How Science Works

For a range of devices, you should be able to:
- describe the intended energy transfers/transformations
- describe the main energy wastages that occur.

Questions

(Grades G-E)

1 How much useful kinetic energy is provided by the engine in the second diagram?

(Grades D-C)

2 Calculate how much heat energy is wasted in the exhaust gases and as heat in the moving parts in the engine in the fourth diagram.

(Grades G-E)

3 Name the **three** ways in which wasted heat energy can be transferred to the environment.

(Grades D-C)

4 Explain why a bath full of warm water is said to be a 'less concentrated' store of energy than a kettle full of hot water.

Energy, work and power

Work and power

- The **work** that a machine does is measured in **joules**.

- The **power** of a machine is measured as the number of joules of work it can do in 1 second. Power is the *rate* of doing work. The unit of power is the **watt** (**W**).

- One joule per second (J/s) = 1 watt of power.

> **Top Tip!**
> Remember to use seconds, not minutes, when calculating power.

G–E

Calculating energy and power

- An object gains **energy** when lifted to a higher position:
  energy gained = work done

- The work done by a **force** is given by: **work done = force × distance moved**

- Raising an object of mass 50 kg by 1 m requires a **force** to lift the object's **weight** (measured in newtons):
  force to overcome weight = mass (kg) × 10 (force of gravity on 1 kg mass)

  potential energy gained = work done = force × distance moved
  = 50 × 10 × 1 = 500 J

- We calculate the power of a machine from:
  power (W) = work done (J) ÷ time taken (s)

- A crane lifts 2000 kg of bricks to a height of 2 m in 2 min:

  work done = force × distance moved = weight × distance moved
  = 2000 × 10 × 2 = 40 000 J
  power = work done ÷ time taken
  = 40 000 ÷ 120 = 333 W

Could you calculate the work you do to climb up a flight of stairs?

D–C

Efficiency

Efficiency

- An **efficient** machine does the maximum work with the minimum of energy input.

- $\text{Efficiency} = \dfrac{\text{useful energy output}}{\text{total energy input}}$

  - Useful energy output is the work a machine does or the energy that a device transforms into useful forms.

> **How Science Works**
> You should be able to: calculate the efficiency of a machine using:
> $\text{efficiency} = \dfrac{\text{useful energy transferred by the machine}}{\text{total energy supplied to the machine}}$

G–E

- A crane has an energy input of 2000 J and does 800 J of useful work. Its efficiency = 400 J per 1000 J.

- The greater the **percentage** of energy that is usefully transformed in a machine, the more efficient a machine is.

- For the petrol engine shown:

  $\text{efficiency} = \dfrac{\text{useful energy output}}{\text{total energy input}} = \dfrac{300}{1000} = 0.3 = 30\%$

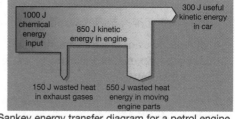

A Sankey energy transfer diagram for a petrol engine.

D–C

Questions

Grades G-E
1 A girl lifts twenty 2 kg bags of sugar from the ground onto a shelf 3 m above the ground.
How much work has she done? (Hint: remember to calculate the *weight* of the bags first.)

Grades D-C
2 It took the girl 1 min. Calculate her power output.

Grades G-E
3 Machine A has an input of 1000 J and a useful energy output of 500 J. Machine B has an input of 2000 J and a useful energy output of 900 J. Which machine is more efficient?

Grades D-C
4 Calculate the percentage efficiency of machine A.

Using energy effectively

Energy saving

- If you cut down on energy use in the home, it saves money and is good for the environment:
  - turn off lights and turn down the temperature of central heating
  - replace old, energy-wasting devices with new, more efficient ones
  - replace incandescent light bulbs with **energy-saver light bulbs**.
- Energy-saver light bulbs cost more than incandescent bulbs but use less than one-fifth of the energy.

Home insulation

- Installing insulation reduces the heat energy lost through ceilings, walls, doors and windows. The table compares savings and costs. **Payback time** is how long it takes for savings to equal installation costs.

Types of insulation.

type of insulation	installation cost (£)	annual saving (£)	payback time (years)
loft insulation	240	60	4
cavity wall insulation	360	60	6
draught-proofing doors and windows	45	15	3
double glazing	2500	25	100

Savings as a result of different types of insulation.

Why use electricity?

Using electricity

- As an energy source in the home, **mains electricity** is instant, convenient, clean, safe and reliable.
- **Electrical energy** is **transformed** into:
  - heat and kinetic energy by a hairdryer
  - light and sound energy by a television.

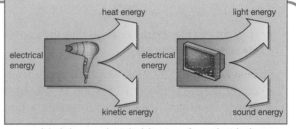

A hairdryer and a television transform electrical energy.

- **Batteries** store and supply a *small* amount of portable **chemical energy** that can be transformed into electrical energy, e.g. for a radio or a mobile phone.

Alternatives to electricity

- The following transform the chemical energy stored in either food or fuel:
  - human, animal and steam power all provide **kinetic energy**
  - various types of biomass (plant material and animal waste) can be burnt to supply **heat** energy for cooking; wax candles or gas or oil lamps can provide **light**.
- **Solar energy** and the **potential energy** in clockwork springs can charge rechargeable batteries. This means that people can use phones, computers and other devices even where there is no electrical supply.

Questions

(Grades G-E)
1 How does putting draught excluders against doors save energy?

(Grades D-C)
2 An ordinary light bulb loses 95% of its energy input as heat. A fluorescent bulb is 20% efficient. Which of these two kinds of bulb gives the most light for the same input of electrical energy?

(Grades G-E)
3 Name the kind of energy that is stored in a battery.

(Grades D-C)
4 Some illuminated signs on country roads are powered by solar and wind energy. Why is this done?

Electricity and heat

Electrical energy and heat

The correct fuse should stop a keyboard from catching fire.

- **Electrical current** from the mains carries **electrical energy** which is **transferred** to **electrical appliances**.

- When an appliance **transforms** electrical energy, some **heat energy** always results. It is often wasted.

- Sometimes, we want the heat energy. For example, the fine wire of a **fuse** gets hot and melts if too much current flows through it. This breaks the circuit and protects other **components**.

G–E

Heat, resistance and current

- A wire heats up because it has **resistance** to the current flowing through it. Thin wires have a greater resistance than thick wires.

- Different metals have different resistances. Copper wire has a relatively low resistance, so current flows easily through it, little electrical energy is transformed into heat, and more is available for circuit components.

- Thickness of wire used in an appliance is chosen according to the current through it. A large current needs a thicker wire. For a telephone it is thin. For an electric cooker it is very thick.

D–C

The cost of electricity

Electrical energy and power rating

- The electric current in the **circuit** of a house *does not get used up*: current in = current out.

- We pay for the **electrical energy** that the current *transfers* to **electrical appliances**.

- **Power rating** is the rate at which an appliance transforms electrical energy. With a power rating of 1 **watt** (**W**), an appliance transforms 1 **joule** (**J**) of electrical energy per second. 1 **kilowatt** (**kW**) = 1000 W. (See also page 52.)

How Science Works

You should be able to:

- compare and contrast the particular advantages and disadvantages of using different electrical devices for a particular application

- use the correct equation to calculate:
  - the total *amount* of energy transferred from the mains
  - the total *cost* of energy transferred from the mains.

G–E

How to calculate energy transfer and its cost

- **i** A 1.2 kW microwave oven is used for 5 min.

  total amount of energy transferred (in J) = power rating (in W) × time (in seconds)

  Total energy transferred from mains supply to microwave oven = 1200 (W) × 300 (s) = 360 000 J

- **ii** Electricity bills charge for kilowatt hours (kWh):

  number of kilowatt hours used = power rating (in kW) × time (in hours)

  Number of kilowatt hours used = 1.2 (kW) × 5/60 (h) = 0.1 kWh (or 0.1 Unit on the electricity bill)

- **iii** The cost of using an appliance is given by:

  total cost = number of kWh used × cost per kWh

  At 8p per kilowatt hour, the cost of using the microwave: = 0.1 × 8 = 0.8p

D–C

Questions

(Grades G-E)

1 Explain why a fuse breaks if too much current flows through it.

(Grades D-C)

2 Explain why a thin wire gets hotter than a thick wire when the same current flows through both of them.

(Grades G-E)

3 A small electric desk fan has a power rating of 50 W. How many joules of electrical energy does it transform every second?

(Grades D-C)

4 If electricity costs 8p per kWh, calculate the cost of running the fan for 30 minutes.

The National Grid

Voltage in the National Grid

G–E

- The **National Grid** is a nationwide network of power stations and cables carrying electricity to homes and businesses.

- Electricity is transmitted through the National Grid at a very **high voltage** (400 000 V) and very **high energy**.

D–C

- At **substations**, **step-down transformers** reduce the voltage to 132 000 V. This is repeated by step-down transformers at smaller substations, so that electricity is at 230 V when it reaches homes and businesses.

> **Top Tip!**
> Very high voltages transmit very large amounts of energy at low current and therefore with relatively small energy losses.

Why the National Grid uses high voltages

- **Power** (in watts) is the energy that an electric current transfers per second:
  power = current × voltage, or **$P = IV$**

- Using very high voltage, the National Grid transmits very high power electricity with a small current.

- Small current means low loss of energy wasted as heat. Thinner, cheaper, copper wires can be used.

- At small substations, the current is split into **parallel**, smaller currents for supply to homes.

Generating electricity

Electric motors and dynamos

G–E

- The blades of a motor-driven fan spin. The **electric motor** transforms **electrical energy** from a battery into **kinetic energy**.

- When the blades are spun by hand and an ammeter replaces the battery, the ammeter records an electric current. Now, kinetic energy is being transformed into electrical energy. The set-up is now a **dynamo**.

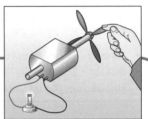

Spinning the blades generates electricity that lights the bulb.

Magnets, current and movement

- An electric current flowing through a wire coil will generate a **magnetic field** that attracts (or repels) a magnet. Either the coil or the magnet will *move*.

  – An *electric current* and a *magnet* together cause *movement*.

D–C

- Moving a magnet into or out of a coil generates an electric current.

- The direction of current as the magnet moves into the coil is reversed when the magnet moves out.

  – A *magnet* and *movement* together cause an *electric current*.

- Wind rotates the windmill blades attached to the wire coil between the magnet poles. This generates a current that the ammeter records. This is how an **electricity generator** works.

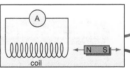

When the magnet moves in or out of the coil, an electric current is generated in the circuit.

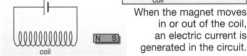

When current flows through the coil, the magnet is attracted to the coil.

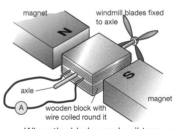

magnet · windmill blades fixed to axle · axle · magnet · wooden block with wire coiled round it

When the blades and coil turn, an electric current is generated.

Questions

Grades G-E

1 Name the device that changes high-voltage electricity into lower voltage electricity before it reaches your home.

Grades D-C

2 Explain why electricity is transmitted at very high voltages in the National Grid.

Grades G-E

3 Explain the difference between an electric motor and a dynamo.

Grades D-C

4 Explain how a magnet and a coil of wire can be used to produce electricity.

Power stations

How a power station generates electricity

- Most British **power stations** burn the **non-renewable**, **fossil fuels** coal, natural gas or oil:
  - heated water becomes steam, which rotates a **turbine**
  - the turbine turns a **generator**, which generates electricity
  - a **step-up transformer** increases the **voltage** to 400 000 V for transmission in the **National Grid**.
- A power station may serve several hundred thousand homes and businesses.

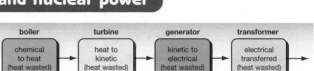

Stages in electricity generation of a coal power station.

Energy changes, efficiency and nuclear power

- Energy transformations in a power station inevitably cause **wasted energy**.
- Combined cycle gas turbine power stations make use of the 'waste' heat to produce steam and drive steam turbines. Efficiency can rise to 50%.
- A **combined heat and power** station reaches 70–80% efficiency if waste energy heats local houses and businesses.
- **Nuclear power stations** are 30% efficient. Uranium-235 undergoes **nuclear fission**, producing vast amounts of heat that power steam turbines. Radioactive plutonium-239 in the waste must be disposed of safely.

boiler	turbine	generator	transformer
chemical to heat (heat wasted)	heat to kinetic (heat wasted)	kinetic to electrical (heat wasted)	electrical transferred (heat wasted)

Energy wasted as heat in a power station. In a coal-burning power station, 65% of energy can be wasted.

Top Tip!

Make sure you know the sequence of stages in a power station and the energy changes at each stage.

Renewable energy

Renewable energy sources

- Some power stations burn household rubbish, and renewable **biomass** (including willow saplings and grass).
- **Wind turbines** transform the **kinetic energy** of moving air to **generate** electricity. They can be grouped into **wind farms**.
- **Hydroelectric** power stations transform the energy of falling water into electrical energy. The water directly turns the turbines as it flows through them. This is very clean, renewable energy.

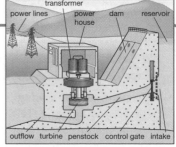

A hydroelectric power station. The transformer steps up the voltage of electricity before transmission along power lines.

- A **tidal power** station uses the energy of water as it moves with the rise and fall of tides. The incoming tide flows through a turbine. The outgoing tide turns the turbine the other way.
- A **wave power** generator uses the energy of wave water inside a tube. A turbine at the top turns when the air above the rising and falling water is forced or sucked through the turbine.
- In a **geothermal power** station, water is pumped down to hot rocks and returns as steam to drive a turbine.

Questions

(Grades G-E)

1 What is the difference between a turbine and a generator?

(Grades D-C)

2 If 65% of the energy in the fuel a power station uses is wasted, what is the efficiency of the power station?

(Grades G-E)

3 Explain why hydroelectric power stations are said to use a 'renewable energy source'.

(Grades D-C)

4 In volcanic areas, such as Iceland, hot water and steam come up to the surface of the ground. How can this be used to generate electricity, and what is the name for this kind of energy source?

Electricity and the environment

Comparing means of electricity generation

- Most means of generating electricity have harmful effects and benefits.

source of electricity	harmful effects	benefits
fossil fuel-burning power station	gases add to global warming, and climate change	currently fuel is cheap and abundant
waste-burning power station	toxic gases may cause cancer, birth defects	reduces need for landfill sites
hydroelectric power station	dams flood valleys, habitats, land, homes	cheap, clean, renewable energy
tidal and wave power station	change water flow patterns; disrupt shipping; destroy wildlife habitats	very low running costs
geothermal power station	might release dangerous gases from deep under the Earth's surface	small station, no deliveries required, so low environmental impact
wind turbine	may kill birds; some people think they are an eyesore	cheaper in remote areas than installing National Grid
nuclear power station	danger of radioactive pollution; problem of nuclear waste disposal	fuel available; no greenhouse gases produced

- To **limit costs**, power companies **locate**:
  - all stations as close to users as possible
  - fossil fuel power stations near roads or rail for fuel access, and a river for cooling water
  - wind turbines in windy areas
  - hydroelectric power stations in hilly landscapes
  - geothermal power stations on areas of easily drilled rock.

How Science Works
You should be able to: compare and contrast the particular advantages and disadvantages of using different energy sources to generate electricity.

Making comparisons

Which type of power generation?

- Fossil fuels produce carbon dioxide which promotes global warming and climate change. They will eventually run out.

- Scientists design **new electricity power stations** that:
  - use fossil fuels more **efficiently** and use a **reliable** energy source
  - use **renewable energy** sources; efficiency is less important
  - are **flexible** enough to meet changing demand.

- **Solar cells** transform the Sun's light energy into electrical energy that can be used for heating, cooling and running many electrical devices and tools.

- Power stations cannot store electrical energy. They have to match output to **demand**.

- Power station planners consider: **capital** (building) **costs**; **operating** (fuel and running) **costs**.

- Most **renewable** energy power stations have low operating costs. Capital costs are high, but improved technologies will reduce them.

- Fossil fuel power stations are cheaper to build, but fuel costs are rising.

- Wind turbines are only **reliable** where winds are regularly over 50 km per hour.

- Hydroelectric power needs reliable rainfall to give a regular flow of water.

- Solar power is practicable where there is year-round sunshine.

Top Tip!
Remember at least one really clear advantage and one really clear disadvantage for each type of power station.

Questions

Grades G-E

1 Explain why people may not want wind turbines near their houses.

Grades D-C

2 Why are oil-fuelled power stations often sited near to a major road?

Grades G-E

3 List **two** reasons why we need to use less fossil fuels to generate electricity.

Grades D-C

4 Explain why hydroelectric power stations are usually situated in mountainous areas of the country.

P1a summary

Heat is energy that flows from a hot object to a colder one.
Temperature indicates how hot something is.

Heat energy can be transferred by **conduction**, **convection** and **thermal radiation**.

Heat energy

Thermal (infrared) radiation is the transfer of heat energy by **electromagnetic waves**.

All materials **absorb** and **emit** thermal radiation.
Dull, dark surfaces do so more than shiny, light surfaces.

Energy **transfer** occurs when the *same* form of energy *moves* from one place to another.
Energy **transformation** occurs when *one* form of energy is *changed* to *another*.

Energy transfer and transformation

Thermal conductors (e.g. metals) transfer he energy easily. **Thermal insulators** (e.g. plastic glass) do not.

Free electrons in metals help to transfer energy between the fixed particles.

Types of energy include: **heat**, **light**, **sound**, **chemical**, **microwaves**, **radio waves**, **kinetic** and **potential**.

The **Law of Conservation of Energy** states:
Energy cannot be created or destroyed. It can only be transferred or transformed from one form to another. The total energy always remains constant.

A **Sankey diagram** shows the **energy input** and **output** of a device.
Wasted energy is usually heat energy.

Energy types and diagrams

Work is measured in joules (J):
work done = force × distance moved

Power is measured in watts (W):
power = work done ÷ time taken

$$\text{efficiency} = \frac{\text{useful energy output}}{\text{total energy input}}$$

Electrical current from the mains carries **electrical energy** which is **transferred** to **electrical appliances**.

The **power rating** of an appliance is the rate at which it transforms electrical energy.

The **cost** of the energy that current transfers to an electrical appliance can be calculated:
- **total amount of energy transferred (J) = power rating (W) × time (s)**
- **number of kilowatt hours (kWh) used = power rating (kW) × time (h)**
- **cost of electricity = number of kWh used × cost per kWh**

The **National Grid** transmits **electricity** around the country at **high voltage** and **low current** to reduce energy losses.

Electricity

A **dynamo** produces electricity when coils of wire rotate inside a **magnetic field**.

Fossil fuels and **biomass** are burned to produce heat.
Nuclear fuels release energy as heat.
The heat is used to produce **steam**:
- the steam turns a **turbine**
- the turbine turns a **generator**
- the generator produces **electricity**.

Renewable energy sources include **wind**, **hydroelectric**, **tidal**, **wave** and **geothermal power**.

All types of **electricity generation** have some **harmful effects** on people or the environment. There are also **limitations** on where they can be used.

Uses of electromagnetic radiation

What is electromagnetic radiation?

- **Electromagnetic radiation** is **energy** carried by **waves** that travel from one place to another. The waves:
  - move in **straight lines**
  - can travel through a **vacuum** (space); some can also travel through matter
  - have a **range** of energies, representing different types of radiation
  - all travel at the **same speed**, the speed of light.

- We group radiations into 'types' according to their effects and uses. We call some types of radiation **rays**.

- The different types of waves have different energies. The **electromagnetic spectrum** shows this.

Types of radiation and their uses

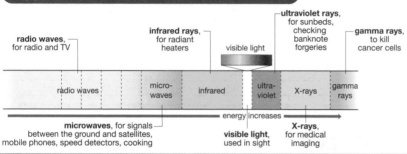

The electromagnetic spectrum.

Electromagnetic spectrum 1

Waves travel while material stays put

- An electromagnetic wave carrying energy moves through material at the speed of light but the material stays where it is.

Top Tip!

The shorter the wavelength, the higher the frequency. The longer the wavelength, the lower the frequency. The higher the frequency, the higher the energy.

Wavelength, frequency and energy

- A wave has a **wavelength**, measured from one crest to the next.

- The shorter the wavelength, the more **energy** it carries.

- The shorter the wave, the greater the number of waves passing a point in one second. This number is the **frequency** of the wave, measured in **hertz (Hz)**. One Hz is one wave per second.

- The diagram shows the wavelengths and frequencies of the waves in the **electromagnetic spectrum**.

lower energy
wavelength

higher energy
wavelength

Wavelength, frequency and energy are linked.

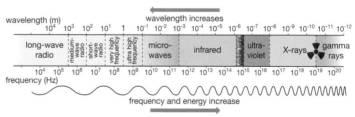

The electromagnetic spectrum with wavelengths and frequencies.

Questions

1 What is carried by an electromagnetic wave?

2 Give **one** use of microwaves, other than cooking.

3 Which electromagnetic waves have the longest wavelength?

4 Which carries more energy – a wave with a short wavelength, or a wave with a long wavelength?

Electromagnetic spectrum 2

Colour, radiation types and sources

- We can see some wavelengths in the electromagnetic spectrum. They are in the **visible range** of wavelengths called **light**. Our eyes absorb energies in this range and send messages to the brain. The brain interprets messages about different groups of wavelengths as different colours.

- Sun, bulbs and lamps **transmit** radiation in the visible range.

- White light contains all colours in the visible spectrum.

- An object that: **reflects** all light radiation looks white; **absorbs** all but red, and reflects red only, looks red; is smooth and shiny and reflects all light is a mirror; absorbs all light radiation looks black.

- Radiation carries energy. A white object absorbs no light energy and a black object absorbs a lot. This is why an illuminated black object is hotter than a white object.

- Types of radiation do not have clear boundaries, even though diagrams show them this way. The table describes radiation types and sources.

wave type	wavelength	sources	detectors
gamma rays	10^{-12} m	radioactive nuclei	Geiger–Müller tube
X-rays	10^{-10} m	X-ray tubes	Geiger–Müller tube
ultraviolet	10^{-7} m	Sun, very hot objects	photographic film, fluorescent chemical
visible light	0.0005 mm	hot objects, Sun, lasers, LEDs	eyes, photographic film
infrared	0.1 mm	warm or hot objects	skin, thermometer
microwaves	1–10 cm	radar, microwave ovens	aerial, mobile phone
radio waves	10–1000+ m	radio transmitters	aerial, TV, radio

Waves and matter

How electromagnetic waves behave

- A mirror **reflects** the infrared radiation of a TV remote control.

- Food in a microwave oven **absorbs** microwave radiation.

- Soft tissue **transmits** X-rays – they pass through it.

- Dark, dull surfaces tend to absorb a large range of radiation.

The mirror reflects the infrared wave and the TV absorbs it.

- The energy from different wavelengths of electromagnetic radiation can be reflected, absorbed or transmitted by an object, depending on its substance and surface.

- To be absorbed, a wave's energy must match the energies that particles of a particular substance can absorb.

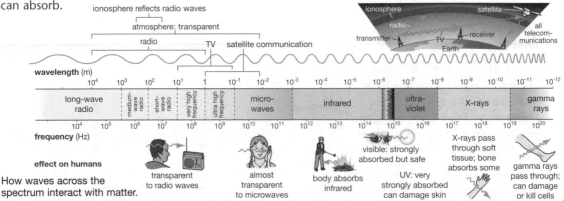

How waves across the spectrum interact with matter.

Questions

Grades G-E
1 What do we call the part of the electromagnetic spectrum that our eyes can detect?

Grades D-C
2 List the different types of electromagnetic radiation, in order of their wavelengths (smallest first).

Grades G-E
3 Why does food inside a microwave oven get hot?

Grades D-C
4 Your mobile phone produces microwaves. Are they absorbed in your head?

Dangers of radiation

Protection from radiation dangers

- Long wavelength radiations (low frequency, low energy) pass harmlessly through humans (see bottom diagram page 60).
- Microwaves cook food and also damage our cells, so ovens start only when the door is closed.
- Strong **infrared** radiation can cause burns if you are too close to a fire.
- Sun-block protects skin from the Sun's **ultraviolet** radiation, which can cause skin cancer.
- Lead aprons and lead-glass screens protect hospital staff from **X-rays** used for imaging and **gamma rays** used to kill cancer cells.

Effect of radiation on living cells

- Radiation can transfer energy to cells and damage them. Effects depend on the type of radiation and size of dose.
- Microwaves in a cooker transfer energy to water molecules, which vibrate more vigorously and heat up.
- Human bodies transmit and absorb infrared radiation. Small doses are safe.
- Molecules in the retina of the eye detect visible light. Too strong a light can damage the retina.
- Ultraviolet radiation, X-rays and gamma radiation are all **ionising** radiations. They can ionise atoms and molecules, especially DNA, in living cells. A cell with damaged DNA can become cancerous.
- Limits for safe doses are set for sun-bed users and hospital patients.

Telecommunications

Fibre optics and digital signals

- The bottom diagram on page 60 shows that radiations from long-wave radio to microwave are used for **telecommunications** through the atmosphere.
- Infrared and visible (**laser**) rays transmit information along **fibre optic cables**, which: replace expensive copper cables used for electrical transmission; transmit signals at the speed of light more clearly (no interference) than electricity; can carry many more messages than wires; allow information to travel curved paths.
- Messages travel as pulses of radiation, called a **digital signal**.

Communications technology

- **Telecommunications** use **geostationary satellites** that stay over the same ground position. The satellites receive and transmit ultra-high frequency and microwave radiations, such as TV signals.

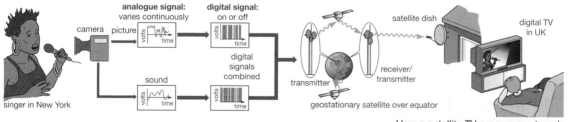

How a satellite TV programme travels.

Questions

Grades G-E
1 What kind of radiation does sun-block protect you from?

Grades D-C
2 How can ionising radiation damage living cells?

Grades G-E
3 What is meant by a 'digital signal'?

Grades D-C
4 Explain what is meant by a 'geostationary satellite'.

Fibre optics: digital signals

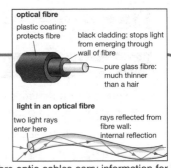

optical fibre
plastic coating: protects fibre
black cladding: stops light from emerging through wall of fibre
pure glass fibre: much thinner than a hair

How light travels along optical fibres

- Long-distance cables for phones, TVs and computers contain **optical fibres**.

- Laser light (infrared and visible) can travel inside a glass fibre. The light is **pulsed** (turned on/off) to produce a **digital signal**.

- An optical fibre is a very fine strand of pure silica glass. Black cladding prevents light from escaping.

- A light pulse in a fibre hits the side of the glass at an **angle of incidence** over 42°, so that the light undergoes **total internal reflection**.

- One fibre can carry several wavelengths of light simultaneously.

- Up to 10 billion pulses travel per second. One fibre can carry over 10 000 phone calls simultaneously.

- Between 12 and 200 optical fibres are bundled together in a cable.

- Light does not heat the cable, and signals in adjacent fibres do not interfere.

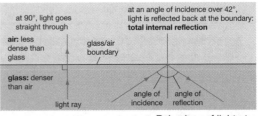

light in an optical fibre
two light rays enter here
rays reflected from fibre wall: internal reflection

Fibre optic cables carry information for phones, cable TV and computers.

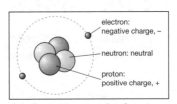

at 90°, light goes straight through
at an angle of incidence over 42°, light is reflected back at the boundary: **total internal reflection**
air: less dense than glass
glass/air boundary
glass: denser than air
light ray
angle of incidence
angle of reflection

Behaviour of light at a glass/air boundary.

Radioactivity

Atoms, isotopes and radiation

- All atoms of the same element have the same number of **protons** in their nucleus. But some of the atoms may have a different number of **neutrons**.

- Different forms of an element are called **isotopes**.

- If an atom is **unstable**, it breaks down and emits **radiation** from its nucleus. It is a **radioactive isotope** of the element.

- You cannot predict when an unstable atom will break down. Breakdown is random.

electron: negative charge, –
neutron: neutral
proton: positive charge, +

A helium atom contains 2 protons, 2 neutrons and 2 electrons.

element	hydrogen	helium	lithium	beryllium	boron	carbon
atomic number	1	2	3	4	5	6
full symbol	^1_1H	^4_2He	^7_3Li	^9_4Be	$^{11}_5B$	$^{12}_6C$

- For an element: all atoms have the same number of **protons**, called the **atomic number**; the number of **neutrons** can vary. Different forms are **isotopes** of the element, with different **mass numbers**. Some are unstable.

- When the unstable **nucleus** of an atom breaks down, it emits **energy**. This breakdown is called **radioactive decay**.

- The energy is emitted as alpha particles, beta particles or gamma rays, depending on the element. It is called **nuclear radiation**.

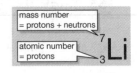

mass number = protons + neutrons
atomic number = protons
^7_3Li

Questions

Grades G-E
1 What is an optical fibre made of?
Grades D-C
2 Explain why light rays do not escape from optical fibres.

Grades G-E
3 An atom has a nucleus containing one proton. Another atom has a nucleus containing one proton and one neutron. Are they atoms of the same element?
Grades D-C
4 Explain why 'nuclear radiation' is given that name.

Alpha, beta and gamma rays 1

Three types of nuclear radiation

- An atom of a radioactive isotope emits **nuclear radiation** as its nucleus breaks up. This happens in one of three ways.

- An **alpha particle** is a helium nucleus. It is large, heavy and cannot go through paper.

- A **beta particle** is an electron. It leaves a nucleus when a neutron changes into a proton. It penetrates aluminium to about 3 mm.

- **Gamma rays** are rays of electromagnetic radiation travelling at the speed of light. Some penetrate lead to a depth of 10 cm.

- The energy carried by types of nuclear radiation varies according to the radiation source. The values in the table are averages.

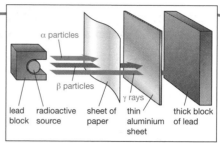

The three radiations and their ability to penetrate materials.

How Science Works

You should be able to: evaluate the possible hazards associated with the use of different types of nuclear radiation.

radiation	alpha (α) particle	beta (β) particle	gamma (γ) rays
what it is; charge	helium nucleus, ^4_2He (large, heavy); 2+	high energy, fast moving electron; e⁻	very high energy/frequency electromagnetic radiation; no charge
range in air	1–5 cm	10–100 cm	very long range
ionising effect	very high	high	low
depth in skin	fraction of a millimetre	0.1 mm	centimetres
damage done to cells	severe, but only if it gets inside the body	moderate	little damage done at low doses, because it is not strongly ionising
uses	smoke detectors, killing cancer cells	radiotracers, medical images	high doses used to destroy cancer cells; as radiotracer to produce images

Background radiation 1

Sources of background radiation

- Very low level, **background radiation** is all around us, in the air and the ground. It includes man-made radiation.

- The Earth's magnetic field and the atmosphere shield us from harmful solar **cosmic rays**.

- Living things have evolved to repair damage to their cells caused by average levels of background radiation.

- Minerals in granite rocks contain atoms that emit **gamma rays** and form **radon** gas, also radioactive. This raises background radiation levels in some parts of the UK. In granite areas, ventilation equipment is installed to clear radon gas from buildings.

- Aeroplanes fly in a weaker magnetic field and thinner atmosphere than at ground level, so there is more cosmic radiation. So aircrew monitor their flight schedules to maintain safe radiation levels.

- Anyone working with man-made radiation (e.g. radiographers and nuclear power station workers) wears a **radiation badge** to monitor exposure.

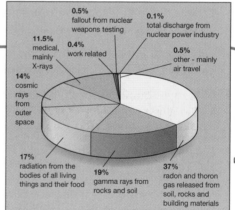

Background radiation chart.

Questions

(Grades G-E)

1 List the **three** kinds of nuclear radiation.

(Grades D-C)

2 Explain why an isotope emitting alpha radiation is very dangerous if you swallow it, but not if you just touch it.

(Grades G-E)

3 Name the **two** sources that produce the greatest amount of background radiation.

(Grades D-C)

4 Explain why people who work near radiation must wear radiation badges.

Half-life

Decay of radioisotopes

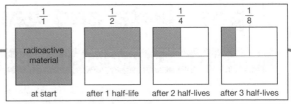

$\frac{1}{1}$	$\frac{1}{2}$	$\frac{1}{4}$	$\frac{1}{8}$
radioactive material			
at start	after 1 half-life	after 2 half-lives	after 3 half-lives

Amount of radioactive isotope left after each half-life.

- Another name for a **radioactive isotope** is **radioisotope**.

- Nuclei of the atoms of radioisotopes undergo the process of **radioactive decay**.

- We cannot predict when this will happen. But we do know how many atoms in a radioisotope sample will decay in a particular time. This is the **activity (count) rate**. Each radioisotope has its unique rate.

- We measure activity as **half-life**.
  The half-life of a radioisotope is the average time it takes for half of its atoms to decay.

- When radioisotopes decay, their nuclei can emit alpha, beta or gamma radiation. A Geiger counter detects and records this radiation.

- As decay continues, fewer unstable nuclei remain and activity rate decreases.

- Scientists identify radioisotopes from their unique rate of decay.

isotope	radiation	half-life
uranium-238	alpha	4.5 billion years
cobalt-60	gamma	5 years
barium-143	beta	12 seconds

- For a sample, the **half-life** is the average time taken:
  - for half the radioactive nuclei to decay, and therefore
  - for the activity (count rate) to decrease to half its starting rate.

How Science Works

You should be able to: evaluate measures that can be taken to reduce exposure to nuclear radiations.

Uses of nuclear radiation

Uses of nuclear radiation

- Gamma rays are used to **irradiate** surgical instruments sealed in plastic, killing bacteria and sterilising them.

- Gamma rays can penetrate the body. In large doses they kill cancer cells. This is called **radiotherapy**.

- Leaks in underground water pipes are located by adding a small amount of radioisotope to the water and detecting a build-up of gamma radiation with a Geiger counter.

- The radioisotope americium-241 in a **smoke detector** emits alpha radiation which ionises air particles. The alarm goes off when smoke particles prevent the ions from reaching the electrodes.

- Paper mills control **paper thickness**. A counter measures the beta radiation that passes through the paper from a radioactive source. If the paper is too thin, the counter reads high and the rollers controlling thickness open a little. If too thick, the counter reads low and the rollers move closer.

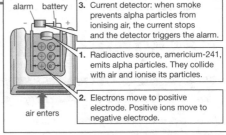

alarm battery
3. Current detector: when smoke prevents alpha particles from ionising air, the current stops and the detector triggers the alarm.
1. Radioactive source, americium-241, emits alpha particles. They collide with air and ionise its particles.
2. Electrons move to positive electrode. Positive ions move to negative electrode.
air enters

How a smoke detector works.

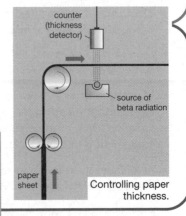

counter (thickness detector)
source of beta radiation
paper sheet
Controlling paper thickness.

How Science Works

You should be able to: evaluate the appropriateness of radioactive sources for particular uses, including as tracers, in terms of the types of radiation emitted and their half-lives.

Questions

(Grades G-E)
1 A piece of a radioactive isotope contains 10 million atoms. After 12 years, 5 million of these atoms have decayed. What is the half-life of the isotope?

(Grades D-C)
2 What can you say about the count rates you could detect from the isotope in question 1, using a Geiger counter, at the start and then 12 years later?

(Grades G-E)
3 How can radiation help us to find leaks in underground pipes?

(Grades D-C)
4 Suggest why beta radiation, not alpha or gamma, is the best to use for checking the thickness of paper.

Safety first

Using radiation safely

- Alpha, beta or gamma radiations can damage cells inside our bodies (e.g. in our lungs).

- Workers using radiation sources follow strict health and safety procedures. They: handle radiation sources carefully, not touching them directly; wear protective gloves and other clothing; wear a badge that monitors the radiation they receive.

- The **energy** of different types of radiation can damage or kill cells by knocking electrons from atoms. This **ionises** the atoms and can disrupt the cells' chemical processes.

- If a cell's DNA is damaged, it can **mutate**. The cell may divide uncontrollably, forming a cancerous tumour.

- Long or intense radiation exposure causes **radiation sickness** and often death, because more cells die than the body can replace.

- There are debates on how to dispose safely of **spent nuclear fuel** which contains radioactive substances.

> **Top Tip!**
> Do not say that radiation ionises 'cells'. You cannot ionise cells, only atoms and molecules.

category of waste	typical waste materials	method of disposal
low-level hazard	protective clothing and used packaging materials	buried in land or at sea
intermediate hazard	irradiated equipment from reactors, cladding materials and reprocessing fluids	storage in concrete warehouses, or deep burial underground
high-level hazard	spent nuclear fuel and other radioactive sources from the reactor	vitrification (sealing in glass blocks), or deep burial

Searching space

Looking into space

> **How Science Works**
> You should be able to: compare and contrast the particular advantages and disadvantages of using different types of telescope on Earth and in space to make observations on and deductions about the Universe.

- **Optical telescopes** use light to form images of objects in space, such as stars and planets.

- The Earth's **atmosphere** makes it difficult to see clear images because: air density and humidity varies and distorts the image; dust pollution reduces the light reaching the telescope. Large ground-based telescopes are therefore built on high mountains in remote, dry places.

- Objects in space transmit radiation in all parts of the **electromagnetic spectrum** (see page 59).

- Ground and satellite telescopes form images of these objects from radio waves through to gamma rays.

- The **Hubble space telescope** was the first to make excellent images in the visible spectrum of objects in space such as the Eagle nebula on page 66.

- A simple optical telescope forms a poor image because it is too short and narrow to gather much light.

- A **reflecting optical telescope** produces far better images.

- The **Hubble space telescope**, a reflecting optical telescope 13.2 m long and 2.4 m in diameter, has formed images further and deeper into space than ever before.

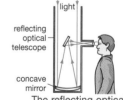

The reflecting optical telescope has a large diameter, and mirrors extend its length.

Questions

1 How can nuclear radiation cause cancer?

2 Explain why disposing safely of spent nuclear fuel is extremely important.

3 Why can a telescope in space produce clearer images of stars and planets than a telescope on the ground?

4 Why did the invention of the reflecting optical telescope allow better images to be seen?

Gravity

Gravity, movement and orbits

- **Gravity** is a 'pull' force acting between *all* objects. A larger **mass** has greater gravity.

- Earth's gravity pulls the smaller Moon, keeping it moving in **orbit**:
  - without gravity, the Moon and Earth would drift apart
  - if the Moon did not orbit, the Moon and Earth would fall together.

- Distant objects have less pull than near objects. The further the **distance** between two objects, the weaker the gravity.

- Gravity and movement keep all planets and other bodies of the **Solar System** in their orbits.

Spacecraft, gravity and space travel

- Launched in 1977, Voyagers 1 and 2 spacecraft have investigated the giant planets and are travelling beyond the Sun's gravity to explore outer space.

- To reduce the fuel load, the spacecraft flew close to Jupiter and Saturn and accelerated past these planets, gaining momentum from their huge gravitational pull.

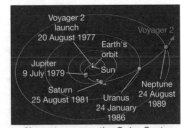

Voyager 2 launch
20 August 1977
Voyager 2
Earth's orbit
Jupiter 9 July 1979
Sun
Saturn 25 August 1981
Uranus 24 January 1986
Neptune 24 August 1989

Voyager 2's route across the Solar System.

Birth of a star

Stars from dust

- Space is not empty, but contains gas and dust particles. In places, these form huge gas clouds called **nebulae**.

- Over millions of years, **gravity** pulls the particles closer, forming a dense ball of gas with a central core.

- Particles forced together can heat up to 15 million °C. Nuclei of atoms fuse together. This is called **nuclear fusion**. It becomes a **chain reaction**, giving out huge amounts of energy. A **star** is formed.

Spectacular gas clouds in the Eagle nebula, taken by the Hubble space telescope.

- The diagram shows the stages as:

  - mass and hence gravity increases, attracting particles that collide and heat up

  - core temperature increases; infrared radiation escapes

  - a protostar is formed; surrounding gas and dust are driven off

  - the protostar collapses; nuclear fusion starts:

immense gas and dust cloud starts to collapse under gravity

cloud gets hotter (invisible core is white hot)

protostar: core heats more, emits particles that drive away close gas and dust

protostar collapses; nuclear fusion starts

size: light takes several years to cross it

infrared radiation

star

Steps in the formation of a star.

nuclear fusion

hydrogen ⟶ helium + **ENERGY**

Questions

(Grades G-E)
1 Write a sentence to say what 'gravity' is.

(Grades D-C)
2 How did gravity help the two Voyager craft to save fuel?

(Grades G-E)
3 What is 'nuclear fusion'?

(Grades D-C)
4 What happens to hydrogen atoms when nuclear fusion takes place?

Formation of the Solar System

Birth and shape of the Solar System

- **Gravity** pulled gas and dust into a huge cloud. A central star heated up, exploded and flung gas outwards.

- The Sun formed and surrounding gases, dust particles and rocks clumped together, forming the planets.

- The inner **rocky planets** are Mercury, Venus, Earth and Mars.

- The outer **gas giants**, Jupiter, Saturn, Uranus and Neptune, are made of gases with a solid core.

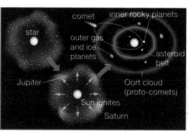

How the Solar System was formed.

- The contracting mass of gas and dust spun into a disc. Dust particles clumped to form rocks.

- The exploding early star shot the gases further out than the rocks.

- The rocks formed the rocky planets, and beyond them, the **asteroid belt**.

- The gases (mostly hydrogen) cooled and condensed into the gas giants.

- On the edge of the Solar System is the Oort cloud of ice and particles, from where many **meteorites** come.

Life and death of a star

Events in a star's life and death

- Astronomers have pieced together the life of stars from their observations.

- A star gradually loses **mass**, as energy and particles in **solar wind**.

- As it loses mass, the star's gravity also gets less, so it expands. It cools as hydrogen fuel runs out.

- A Sun-sized star eventually becomes a **red giant** which collapses under gravity to form a **white dwarf**, then a **black dwarf**.

- A larger star becomes a **red supergiant** which collapses and explodes in a **supernova**. The small core becomes a **neutron star**. This may become a **pulsar**, and, if large enough, a **black hole**.

- In a stable star, two factors are in balance:
  - huge energy and streams of particles from **nuclear fusion** reactions produce an *outward* force called **radiation pressure**
  - the mass of gas exerts an *inward* **force of gravity**.

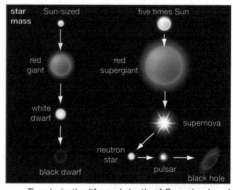

Events in the life and death of Sun-sized and larger stars.

- These become unbalanced when hydrogen in the core runs out. Hydrogen fusion continues further out and the star expands. It continues to lose mass. The cooling core contracts, and the star collapses.
  - The star's **mass** (see diagram) determines the **speed** of collapse and the objects that may then form.

Questions

(Grades G-E)
1 Which are the **four** gas giants?

(Grades D-C)
2 Explain why the gas giants are the planets furthest from the Sun.

(Grades G-E)
3 What will eventually happen to our Sun?

(Grades D-C)
4 Explain why a star collapses when its hydrogen fuel runs out.

In the beginning

The Big Bang

- **Galaxies** are moving apart at tremendous speeds. It is thought that:
  - there was once a huge amount of energy at a single point
  - the energy expanded violently in the **Big Bang**
  - within seconds, the temperature cooled, and energy began changing into **protons**, **neutrons** and **electrons**.
- Over billions of years, galaxies containing stars and planets have formed.

Energy becomes matter

- Following the Big Bang, the Universe expanded and cooled. Energy became matter, including quarks. Cooling quarks formed protons, neutrons and electrons.

- Hydrogen, the first element, formed. It made up the first gas clouds. These were the birthplace of stars, in which nuclear fusion produced the other elements.

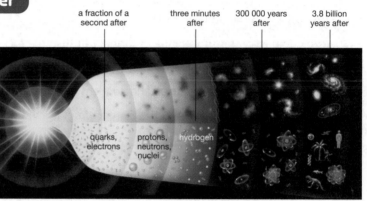

a fraction of a second after | three minutes after | 300 000 years after | 3.8 billion years after

quarks, electrons | protons, neutrons, nuclei | hydrogen

The different stages of the Big Bang.

The expanding Universe

Red shift and an expanding Universe

- Light from a stationary source is at a particular **wavelength**.
- If the light source is moving away from an observer, the same wavelength *appears* longer, shifting towards the red part of the spectrum. The difference between actual and apparent wavelengths is the **red shift**.
- Galaxies are moving apart. Their red shift tells us that the further away, the faster they recede.

The redder the galaxy, the faster it is travelling away, and therefore the greater its distance from Earth.

- A **spectroscope** splits light into bands of different colours (wavelengths), forming a **spectral pattern**.
- When heated, every element emits light which forms a unique spectral pattern.
- Edwin Hubble looked at the spectral pattern for hydrogen in different galaxies. From the red shifts, he worked out how fast each galaxy was receding, and how far away it was.

The spectral pattern for hydrogen.

- George Gamow worked back to a single point as the place where the galaxies started from. This and the red shift support the Big Bang theory.

Questions

1 According to the Big Bang theory, what happened at the very beginning of the Universe?

2 Which was the first element that was formed by the Big Bang?

3 If something is moving away from us, the wavelength of the light coming to us appears to be longer. Rewrite this sentence, stating how the appearance of the *frequency* of the light will change.

4 Explain how the red shift provides evidence for the Big Bang.

P1b summary

Electromagnetic radiation is **energy** carried by **waves** that:
- travel in **straight lines**
- can travel through a **vacuum**
- have a range of energies
- travel at the **speed of light**.

Radiation can be **reflected**, **absorbed** or **transmitted** by an object depending on its substance and surface.

Wavelength is measured from one **crest** to another.
Frequency is the number of waves passing a point in a second, measured in **hertz** (Hz).

Optical fibres use pulsed laser light to produce a **digital** signal.

Electromagnetic radiation

The **electromagnetic spectrum** is divided into regions to show the energies of each particular type of wave.
From low frequency and energy (longest wavelength) to high frequency and energy (shortest wavelength) they are:
- radio waves
- microwaves
- infrared
- visible light
- ultraviolet
- X-rays
- gamma rays.

Radiation can **damage** living cells. **Ionising** radiation (ultraviolet, X-rays and gamma rays) can ionise atoms, including in **DNA**, and can cause cancer.

Radiation has many uses in **communication** systems including: radio, TV, satellites, cable and mobile phone networks.

Radioactive substances have **unstable atoms** that emit radiation.
Three forms are: **alpha particles**, **beta particles** and **gamma rays**.
Each has its own uses and hazards.

Radioactivity

Background radiation is all around us. **Granite** rocks can emit **gamma rays** and form radioactive **radon** gas.

The **half-life** of a **radioisotope** is the average time it takes for half of its atoms to **decay**.

Reflecting optical telescopes built on high ground in dry places and **space telescopes** (e.g. the **Hubble** telescope) form the best images of space as they reduce or avoid **atmospheric distortion** of the image.

Gravity and **movement** keep all planets and other bodies in the **Solar System** in orbit.

A **star** is born when the particles in a **nebula** are pulled together by **gravity**, heat up and form a **protostar**. This collapses and **nuclear fusion** starts:

$$\text{hydrogen} \xrightarrow{\text{nuclear fusion}} \text{helium} + \text{ENERGY}$$

The gases, dust particles and rocks surrounding the **Sun** formed the inner **rocky planets**, the outer **gas planets**, the **asteroid belt** and the outer **Oort cloud** of the **Solar System**.

The Universe

A star's **mass** determines the **speed** of its **collapse** and the objects that may form.
A sun-sized star becomes:
  red giant → white dwarf → black dwarf
A larger star becomes:
  red supergiant → supernova → neutron star → pulsar → black hole.

Current evidence suggests that: the **Universe** is **expanding**; a huge amount of **energy** at a single point expanded violently in the **Big Bang**; energy became **matter**; **hydrogen** was formed into gas clouds; **stars** were born; and **nuclear fusion** produced the other **elements**.

Red shift indicates that:
- **galaxies** are moving apart
- the **further** away a galaxy is, the **faster** it recedes.

Cells

Animal and plant cells

- The top diagram shows the structure of the **cells** of **animals**, including humans.

- A typical **plant** cell includes the parts shown in the animal cell. It is also enclosed in a cell wall and may contain chloroplasts and a permanent vacuole.

Top Tip!

Be very careful to use the terms cell *wall* and cell *membrane* correctly.

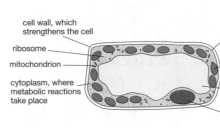

cell membrane, which controls the passage of substances in and out of the cell

cytoplasm, where metabolic reactions controlled by enzymes take place

mitochondrion, where energy is released in respiration

ribosomes, where protein synthesis takes place

nucleus, which controls the activities of the cell

Detail of an animal cell.

cell wall, which strengthens the cell

ribosome

mitochondrion

cytoplasm, where metabolic reactions take place

cell membrane, which controls the passage of substances in and out of the cell

chloroplast, which absorbs light energy to make food

permanent vacuole, filled with cell sap

nucleus, which controls the activities of the cell

A plant cell.

G–E

Cell organelles

- The parts inside a cell are called **organelles**. Each organelle has a particular function.

- Energy is released in respiration in the **mitochondria**.

- Protein synthesis occurs in **ribosomes**.

How Science Works

You should be able to: relate the structure of different types of cells to their function in a tissue or an organ.

D–C

Specialised cells

Different cells perform different functions

- All the functions of a single-celled organism are carried out by the one **cell**.

- A larger, multicelled organism contains different tissues and organs. The cells of each tissue or organ have a structure that is specialised for their **function**.

- The diagram on the right shows how a sperm cell is adapted for its function.

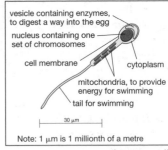

vesicle containing enzymes, to digest a way into the egg

nucleus containing one set of chromosomes

cell membrane

cytoplasm

mitochondria, to provide energy for swimming

tail for swimming

30 μm

Note: 1 μm is 1 millionth of a metre

A sperm cell.

G–E

- Here are two more examples of specialised cells in the human body.

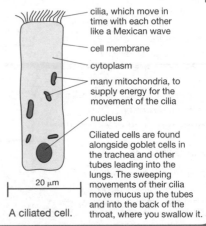

Goblet cells are found in the lining of the alimentary canal, and in the tubes leading down to the lungs. They make mucus, which helps food to slide easily through the alimentary canal, and helps to stop bacteria getting down into your lungs.

mucus that has been made by the cell

cell membrane

cytoplasm

nucleus

20 μm

A goblet cell.

cilia, which move in time with each other like a Mexican wave

cell membrane

cytoplasm

many mitochondria, to supply energy for the movement of the cilia

nucleus

Ciliated cells are found alongside goblet cells in the trachea and other tubes leading into the lungs. The sweeping movements of their cilia move mucus up the tubes and into the back of the throat, where you swallow it.

20 μm

A ciliated cell.

D–C

Questions

Grades G-E

1 Name **two** structures in a plant cell that are not in animal cells.

Grades D-C

2 Name the organelle where protein synthesis takes place.

Grades G-E

3 State **two** ways in which a sperm cell is adapted for its function.

Grades D-C

4 Where are goblet cells found and what do they do?

Diffusion 1

Moving particles

* **Diffusion** is the spreading of particles of a gas, or of a substance dissolved in a solution, from a region where the particles are at a higher **concentration** to where they are at a lower concentration.

Diffusion and temperature

* Water and sugar molecules in the solution move around in all directions. The sugar **concentration** is higher at the bottom of the beaker than at the top.

* By diffusion, there is a **net movement** of sugar molecules from a more concentrated to a less concentrated region since, on average, there are more to move to the less concentrated region than in the other direction. Eventually, sugar and water molecules are evenly spread throughout the solution.

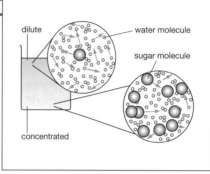

Diffusion in a beaker of sugar solution.

* Particles move faster at higher **temperatures**. Therefore, a temperature rise speeds up diffusion.

Diffusion 2

Diffusion in cells and jelly

* Cells need **oxygen** for **respiration** according to the equation:

glucose + oxygen → carbon dioxide + water

* Respiration uses up oxygen, so oxygen concentration inside the cell is lower than outside. This causes oxygen to diffuse into the cell through the cell membrane.

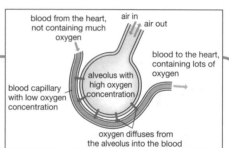

Gas exchange in an alveolus. Why does each lung have millions of alveoli?

* The faster the cell respires, the faster it uses up oxygen, increasing the difference between oxygen concentrations inside and out. The greater the concentration difference, the faster oxygen diffuses into the cell.

* **Alveoli** in the lungs allow oxygen to diffuse rapidly into the bloodstream for transport to cells around the body.

* **Antibiotics** kill bacteria and can diffuse through **agar jelly**.

* The agar contains bacteria, and each disc contains a different concentration of the same antibiotic. The table shows results for bacterial growth after two days.

disc	concentration of antibiotic (arbitrary units)	diameter of clear area (mm)
A	0	0
B	1	4
C	5	6
D	10	9

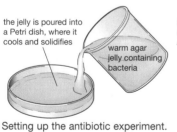

the jelly is poured into a Petri dish, where it cools and solidifies

warm agar jelly containing bacteria

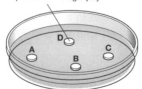

little discs of filter paper, each soaked in a different concentration of antibiotic and placed on the agar jelly

Setting up the antibiotic experiment.

bacteria growing on the agar jelly

The dish after two days.

clear area where no bacteria can grow

Questions

1 Which of these contain particles that can diffuse: solid, solution, gas?

2 How does a decrease in temperature affect the rate of diffusion?

3 Which gas diffuses into a cell when it is respiring?

4 What do the results of the experiment above tell you about how a difference in concentration affects the rate of diffusion?

Osmosis 1

Demonstrating osmosis

Top Tip!
Osmosis is just a special kind of diffusion – the diffusion of water through a partially permeable membrane.

- A **partially permeable membrane** allows small molecules like water to **diffuse** through it, but blocks larger molecules.

- If a partially permeable membrane separates a dilute sugar solution from a concentrated solution, water molecules diffuse from the dilute solution through the membrane into the concentrated solution (see top diagram). This diffusion of water molecules only is called **osmosis**.

- In the bottom diagram, *water* molecules are at a *higher* concentration in the dilute sugar solution, and move to where water molecules are at a *lower* concentration in the concentrated sugar solution.

- In the experiment, all the water and sugar molecules move and bump into each other and the membrane.

- Only water molecules can pass through the holes in the membrane.

- More water molecules from the dilute sugar solution will cross the membrane into the concentrated sugar solution than in the reverse direction.

- Now there is less water in the dilute sugar solution and more in the concentrated solution than at the start of the experiment.

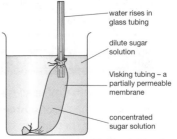

Using Visking tubing to demonstrate osmosis.

Explaining osmosis.

Osmosis 2

Osmosis and animal cells

- The **cytoplasm** of an animal cell is a fairly concentrated solution. The **cell membrane** is **partially permeable**: water can diffuse through it.

- The diagrams show the movement of water by osmosis: *into* an animal cell from distilled water (highest water concentration), the cell bursts; *out* of an animal cell into a more concentrated solution, the cell collapses.

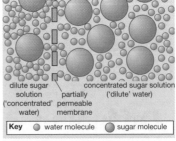

Water moves into this animal cell by osmosis.

Water moving out.

Osmosis and plant cells

- As well as a partially permeable cell membrane, a plant cell has a strong, fully permeable **cell wall** that limits expansion of the cell.

- Therefore, only a limited amount of water can enter the cell by osmosis, and the cell does not burst.

- A plant cell will lose water by osmosis when placed in a solution more concentrated than the cytoplasm. The **cell membrane** may pull away from the cell wall. If it tears, the cell dies.

How could you try to revive this cell?

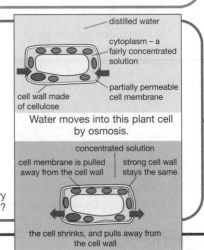

Water moves into this plant cell by osmosis.

Questions

Grades G-E
1 Name the substance that diffuses across a partially permeable membrane during osmosis.

Grades D-C
2 Which contains more water molecules – a dilute solution or a concentrated solution?

Grades G-E
3 Which part of a cell is a partially permeable membrane?

Grades D-C
4 Name the structure that stops a plant cell bursting when it absorbs a lot of water.

Photosynthesis

Making food and using energy

- In **photosynthesis**, the green plant pigment, **chlorophyll**, traps energy from sunlight. The energy is used to convert **carbon dioxide** and water into **glucose**:

    carbon dioxide + water (+ light energy) → glucose + oxygen

- Glucose is a **sugar**, a kind of **carbohydrate**. Plants can use glucose to make other kinds of food, such as **starch**, fats and proteins.

- The by-product of photosynthesis is **oxygen**, which plants release.

- In photosynthesis, the **energy** from light becomes **chemical energy** stored in glucose molecules.

- This energy can be unlocked again when organisms break down glucose in respiration:

    glucose + oxygen → carbon dioxide + water + energy

- Our cells use this energy for warmth, growth and movement.

- Plants also respire, and use the energy to grow and produce flowers and fruit.

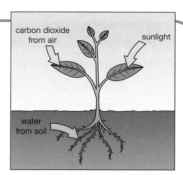

This plant is photosynthesising.

Top Tip!

Remember that plants respire and photosynthesise during the day, and just respire at night.

Leaves

Leaves: adaptations for photosynthesis

- The diagram shows how a **leaf** is **adapted for photosynthesis**.

- Photosynthesis uses up carbon dioxide and its concentration in the leaf becomes less than in air, so more carbon dioxide **diffuses** into the leaf through the **stomata**.

- Photosynthesis produces oxygen which becomes more concentrated in the leaf than outside it, so this gas diffuses out through the stomata into the air.

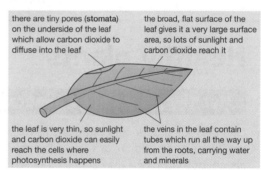

there are tiny pores (stomata) on the underside of the leaf which allow carbon dioxide to diffuse into the leaf

the broad, flat surface of the leaf gives it a very large surface area, so lots of sunlight and carbon dioxide reach it

the leaf is very thin, so sunlight and carbon dioxide can easily reach the cells where photosynthesis happens

the veins in the leaf contain tubes which run all the way up from the roots, carrying water and minerals

How a leaf is adapted for photosynthesis.

- Plant cells adapted for photosynthesis have chloroplasts, which contain chlorophyll to absorb the light energy used to make glucose.

- Glucose can be used immediately to release energy in respiration. Cells store excess glucose as starch.

- The **tissue** best adapted for photosynthesis is the palisade mesophyll.

- Spongy mesophyll and the cells of stomata are adapted for gas exchange.

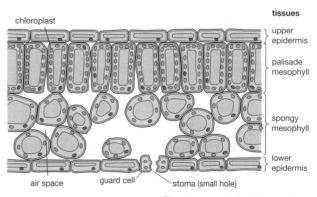

The tissues of a leaf in section.

Questions

1 Write down the word equation for photosynthesis.

2 Name the process in which cells get energy from glucose.

3 Give **two** ways in which being thin helps a leaf to photosynthesise.

4 How does the position of palisade cells in a leaf help them to photosynthesise?

Limiting factors

Speeding up photosynthesis

How Science Works

You should be able to:
- interpret data showing how factors affect the rate of photosynthesis
- evaluate the benefits of artificially manipulating the environment in which plants are grown.

- **Photosynthesis** speeds up when:
  - the **intensity** (brightness) of the **light** increases
  - the **concentration** of **carbon dioxide** increases
  - the **temperature** increases, as happens through spring and summer.

- These factors can be controlled in glasshouses.

Limiting factors for photosynthesis

- The upper graph shows that no photosynthesis occurs at zero **light intensity**. As light intensity gradually increases, the **rate** of photosynthesis increases. In region A, the low light intensity is **limiting** the rate of photosynthesis – it is a **limiting factor**.

- Beyond point B, any further increase in light intensity does not affect the rate of photosynthesis.

- The lower graph shows that, in region A, **carbon dioxide concentration** is a limiting factor for photosynthesis.

- Within certain values, **temperature** also limits the rate of photosynthesis.

- Light intensity, carbon dioxide concentration and temperature are all limiting factors for the rate of photosynthesis, which they can affect separately or together.

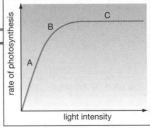

How light intensity affects the rate of photosynthesis.

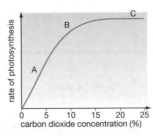

How carbon dioxide concentration affects the rate of photosynthesis.

Healthy plants

Healthy plants need mineral salts

- Using carbon dioxide and water, plants make glucose and other sugars, starch, fats and cellulose. All these are **carbohydrates** – they contain the elements carbon, hydrogen and oxygen only.

| | elements it contains | | | | |
substance	C	H	O	N	Mg
carbohydrate	✓	✓	✓		
fat	✓	✓	✓		
protein	✓	✓	✓	✓	
chlorophyll	✓	✓	✓	✓	✓

- To make proteins and chlorophyll from carbohydrates, plants need other elements. They exist in ions of **mineral salts** in the soil – **nitrogen** as **nitrate ions** and **magnesium** as **magnesium ions**. Plants absorb these ions through their roots.

- For healthy growth, plants need mineral ions.

- If a plant is **deficient** in (lacks) an ion it needs, it will not grow properly.

- A plant deficient in **nitrate ions** cannot make enough amino acids and therefore proteins. Its growth is stunted.

- Each chlorophyll molecule contains a **magnesium** atom. A plant deficient in **magnesium ions** cannot produce enough chlorophyll and its leaves are yellow.

A healthy potato plant.

A potato plant short of nitrate ions.

A potato plant short of magnesium ions.

Questions

Grades G-E
1 List **three** factors that could increase the rate of photosynthesis.

Grades D-C
2 In the light intensity graph, why does the line stay flat in region C?

Grades G-E
3 Name a compound found in plants that contains magnesium ions.

Grades D-C
4 How do nitrate ions help to make proteins?

Food chains

A lion with the carcass of an antelope, and vultures in the background.

Energy flow

- A **food chain** shows how energy (originally from sunlight) passes from one organism to another.

- In the food chain: grass → antelope → lion
  - grass is a **producer**. It uses light energy to grow, providing nutrients that the next organism consumes. Food chains always start with a producer.
  - the antelope is the **primary** (first) **consumer** and is a **herbivore**: it eats plants.
  - the lion is a **secondary consumer** and is a **carnivore**: it eats other animals. It is also a **predator**, and the animals it kills are its **prey**.

Food webs and energy wastage

- Food chains can interact to form a **food web**.

- A leaf uses only a little of the solar energy reaching it to store in cell substances. The rest is reflected or passes through the leaf without reaching the **chlorophyll** or it is the wrong wavelength (colour). This energy is wasted.

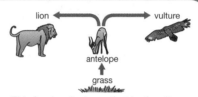

This food web shows that both vultures and lions feed on antelope.

Biomass

How Science Works

You should be able to: interpret pyramids of biomass and construct them from appropriate information.

Pyramids of biomass

- A mass of living material is called **biomass**.

- The biomass of grass that an antelope eats greatly exceeds the antelope's biomass. The biomass of the antelopes a lion eats greatly exceeds its biomass. Also, antelopes do not eat entire plants or lions entire antelopes, yet biomass includes the uneaten parts.

- A **pyramid of biomass** represents the mass at each step in a food chain. In each step, the biomass is less than in the step before.

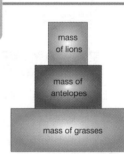

mass of lions

mass of antelopes

mass of grasses

A pyramid of biomass.

Energy losses

- The biomass reduces along a food chain because not all the energy from one organism is transferred to the next.

energy lost as heat from respiration

energy in grass → energy in antelope → energy in lion

How energy is lost in this food chain.

energy lost as waste

- The diagram shows that energy:
  - is transferred to tissues and organs as the antelope and lion grow
  - is lost as heat in respiration and as waste.

- To survive in cold surroundings, mammals and birds use **respiration** to maintain a constant body temperature. They are **homeotherms** (meaning 'same temperature').

- Snakes, frogs and fish stay at the temperature of their surroundings. The biomass of the food they need is less than for homeotherms.

Questions

Grades G-E

1 What do the arrows in a food chain show?

Grades D-C

2 List **three** reasons why plants do not use all the energy in the sunlight where they grow.

Grades G-E

3 What is biomass?

Grades D-C

4 Why do mammals need more food than reptiles of the same size?

Food production

Shortening food chains

It would be more energy-efficient for us to eat the maize rather than eat the cattle.

- Biomass gets less along **food chains** because **energy** is lost at each stage.

- The shorter the food chain, the smaller the energy loss.

- In farming, reducing energy loss increases the **efficiency** of **food production**.

- An inefficient food to produce is salmon: plankton → small fish → bigger fish → salmon → humans

- An efficient food to produce is maize: maize plants → humans

G–E

Energy-saving methods of food production

- Farmers of poultry and livestock find ways to reduce energy losses:
  - birds and animals are housed in barns to reduce their movement
  - barns are warmed to reduce heat loss in respiration.

- The energy in animal feed is then converted efficiently into eggs, meat and milk.

- In other cultures, animals have other uses that affect their living conditions:
  - pigs wander around eating discarded food scraps
  - cattle pull ploughs in fields.

D–C

How Science Works

You should be able to:
- evaluate the positive and negative effects of managing food production and distribution
- recognise that practical solutions to human needs may require compromise between competing priorities.

The cost of good food

Food miles and animal welfare

- Food production costs include transport costs.

- To save costs, supermarket food is carried in container lorries from huge central distribution warehouses.

- Transport costs may be less for locally produced food.

- Cheap food may result from poor **animal welfare**, including bad housing and poor quality feed.

- The minimum welfare standards for rearing animals in Britain may not apply to farming overseas. Therefore, food from overseas is often cheaper.

- Humans need food, but cheap food can mean poor animal welfare.

G–E

Three means of egg production

Battery hens.

- **Battery hens** are cramped in small cages in heated buildings. Automated systems provide their food and water.

- **Barn hens** are kept in heated buildings, but are free to walk around and have perches.

- **Free-range hens** are indoors at night but are free to walk around a fenced area outside during the day.

D–C

Questions

Grades G-E

1 Why is it more energy-efficient to eat bread than to eat beef?

Grades D-C

2 How can keeping animals warm reduce a farmer's costs?

Grades G-E

3 How do supermarkets reduce transport costs for the food they sell?

Grades D-C

4 Why are free-range eggs more expensive than eggs from battery hens?

Death and decay

Decay

- All plants and animals eventually die.

- **Microorganisms**, including bacteria and fungi, use **enzymes** to break down (**digest**) dead and waste plant and animal materials. This is the **decay process**.

- In order to break down these materials, microorganisms need oxygen for **aerobic respiration**.

Speeding or slowing decay

- The growth and decay action of microorganisms are affected by temperature, moisture levels and oxygen levels.

- Microorganisms function most rapidly at **warm temperatures**, typically between 25 and 45 °C. (High temperatures kill microorganisms, but the spores of some survive.)

- To grow and cause decay, microorganisms need moisture.

- To be active, many microorganisms need oxygen for aerobic respiration.

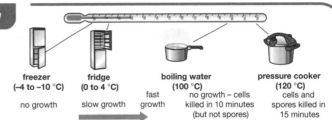

freezer (−4 to −10 °C)	fridge (0 to 4 °C)	boiling water (100 °C)	pressure cooker (120 °C)
no growth	slow growth	fast growth	no growth – cells killed in 10 minutes (but not spores)	cells and spores killed in 15 minutes

How temperature affects the activity of microorganisms.

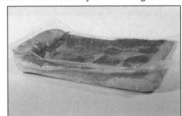

There is no water in this dried food (left), and there is no air in this vacuum-packed food (right), so microorganisms cannot grow.

Cycles

Recycling

- Without **decay**:
  - dead and waste plant and animal materials would pile up
  - nutrients for plants would be locked up in these materials and the soil would be less fertile.

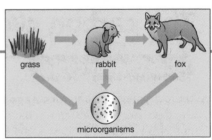

grass → rabbit → fox

microorganisms

A food chain with decay microorganisms.

- Decay by **microorganisms** enables substances taken from the environment to be returned to it.

- **Detritus feeders** are animals that eat dead bodies and waste material, for example earthworms.

- Within a **community** of organisms, materials are removed from the environment and also returned to it – they are **cycled**. If removal processes and return processes are balanced, the community is stable.

- Microorganisms continually break down dead parts of plants and the waste products of animals.

- The diagram shows that microorganisms break down all organisms in a **food chain** when they die.

- The processes of removing materials from the environment and returning them continue all the time, as the diagram shows.

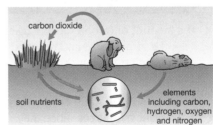

carbon dioxide

soil nutrients

elements including carbon, hydrogen, oxygen and nitrogen

This diagram shows how materials are recycled in an ecosystem.

Questions

Grades G-E

1 How do microorganisms cause decay?

Grades D-C

2 Why does decay happen more quickly when it is warmer?

Grades G-E

3 Give an example of a detritus feeder.

Grades D-C

4 What kind of organisms are fed on by decay microorganisms?

The carbon cycle 1

The importance of carbon

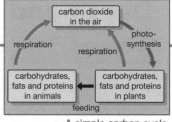

A simple carbon cycle.

- Carbon starts in the **carbon dioxide** of the air. In **photosynthesis**, green plants use carbon dioxide to make glucose:

$$\text{carbon dioxide} + \text{water} \xrightarrow{\text{light energy}} \text{glucose (a carbohydrate)} + \text{oxygen}$$

- For growth, plant cells make fats, proteins and carbohydrates from glucose. All these contain carbon.

- Plants are eaten by animals, and some animals eat each other.

- Plant and animal cells also use glucose in **respiration** to produce carbon dioxide which enters the air, and to release usable energy:

$$\text{glucose} + \text{oxygen} \rightarrow \text{carbon dioxide} + \text{water} + \text{energy}$$

- This constant cycling of carbon is called the **carbon cycle**.

G–E

A Martian greenhouse

- Life on Earth is adapted to live in Earth's atmosphere. About the same amount of carbon dioxide that plants remove from the atmosphere is returned when organisms respire.

- Mars is cold and has a very thin atmosphere, mostly of carbon dioxide.

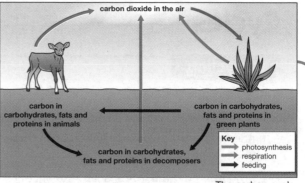

Why would plants not be able to grow in the Martian atmosphere?

- Scientists suggest that they could grow plants on Mars inside huge transparent glasshouses. There would be natural light and carbon dioxide. As plants photosynthesised, oxygen would gradually build up to a level in which visitors from Earth could survive.

D–C

The carbon cycle 2

The carbon cycle

- Microorganisms that are **decomposers**, and the **detritus feeders**, return the plant nutrients in dead organisms to the soil. When decomposers respire, they release carbon dioxide. In this way, all the energy originally captured by green plants has been transferred.

The carbon cycle.

G–E

- The diagram shows how animals, plants and decomposers all interact in the carbon cycle.

Do trees store carbon?

The numbers = mass of carbon taken in or given out per year, in each square metre of forest.

- The amount of carbon dioxide taken from the atmosphere and returned to it is becoming imbalanced: its concentration is increasing in air.

- A way to remove some carbon may be to plant more trees, since they take in carbon dioxide in photosynthesis.

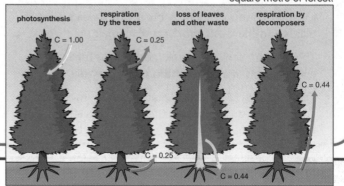

D–C

Questions

Grades G-E

1 Name **three** types of compound in a plant that contain carbon.

Grades D-C

2 Why would we need plants if we lived on Mars?

Grades G-E

3 How do decomposers return carbon dioxide to the air?

Grades D-C

4 How do trees help to remove carbon dioxide from the air?

B2a summary

Cells

Animal cells have a cell membrane, cytoplasm and a nucleus.
In addition, **plant cells** may also have a permanent vacuole, chloroplasts and a cell wall.

In multicellular organisms, different cells are **specialised** to perform different functions.

Particles of gas, or substances dissolved in a solution, move by **diffusion**. Oxygen and carbon dioxide diffuse into and out of cells.

Osmosis is the special diffusion of water molecules through a partially permeable membrane.

Photosynthesis

Green plants use **chlorophyll** to trap **energy from sunlight** to perform **photosynthesis**:
carbon dioxide + water → glucose + oxygen

Leaves are **adapted** for photosynthesis by being very thin and having stomata, a broad, flat surface and veins.

Glucose can be used in **respiration**:
glucose + oxygen → carbon dioxide + water + ENERGY

The **rate** of photosynthesis is affected by:
– light intensity
– carbon dioxide concentration
– temperature.
These can all be **limiting factors**.

Ions of **mineral salts** in the soil can be used by plants to make proteins or chlorophyll.

Lack of a particular mineral ion results in a **deficiency symptom** in a plant.

Food chains and cycles

Food chains show the feeding relationships of organisms.

Energy passes along food chains, but some energy is lost at each transfer.

The shorter the food chain, the smaller the energy loss.

Reducing energy loss increases the **efficiency of food production**.

In **decay**, **microorganisms** use **enzymes** to break down waste materials.

Decay is affected by the levels of:
– temperature
– moisture
– oxygen.

Decomposers and **detritus feeders** feed on the organisms in the food chain, recycling materials, e.g. carbon.

In the **carbon cycle**, carbon is removed from the air as carbon dioxide during photosynthesis and returned to it by respiration.

Enzymes – biological catalysts

Enzyme action, heat and pH

- **Enzymes** are **biological catalysts** that speed up chemical reactions and are not used up. They are **protein molecules**, affected by **temperature** and **pH**. Each controls one type of reaction.

- Enzymes catalyse the reactions in **respiration**, **protein synthesis** and **photosynthesis**.

- An enzyme molecule is a long, folded chain of **amino acids**. It has a hollow **active site** where its **substrate** molecule fits and reacts.

- High **temperatures** and extreme **pHs** (beyond a narrow range) change the *shape* of the active site. Then the substrate cannot fit or react.

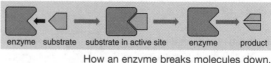
enzyme substrate substrate in active site enzyme product

How an enzyme breaks molecules down.

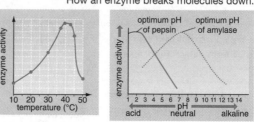

Graphs showing the effect on enzyme activity of temperature (left) and of pH (right).

- Some enzymes (e.g. digestive enzymes) *break large* molecules into smaller ones. Others *join small* molecules to make larger ones (e.g. enzymes joining amino acids to form proteins).

- The temperature at which an enzyme works best is its **optimum temperature**. In humans, this is 37 °C. Cooling enzymes **inactivates** them. Overheating **denatures** (destroys) them.

- Each enzyme has an optimum **pH**. Extreme pHs denature enzymes.

Top Tip!

Do not say that enzymes are 'killed' by high temperatures. They are just chemicals, not living things. They are *denatured*.

Enzymes and digestion

Digestive enzymes

Where enzymes of the human digestive system are produced.

- **Digestive enzymes** are secreted into the gut cavity by specialised cells in glands. They catalyse the splitting of large, insoluble food molecules into smaller ones that pass through villi in the small intestine, into the bloodstream and reach all parts of the body.

- The pancreas produces all three groups of digestive enzymes: amylase, protease enzymes and lipase enzymes.

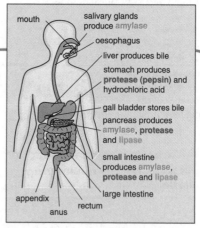
mouth — salivary glands produce amylase
oesophagus
liver produces bile
stomach produces **protease (pepsin)** and hydrochloric acid
gall bladder stores bile
pancreas produces amylase, **protease** and lipase
small intestine produces amylase, **protease** and lipase
large intestine
appendix
rectum
anus

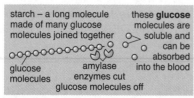

Digestion by salivary amylase of starch in the mouth at pH6.5–7.5.

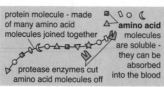

Digestion by a protease (pepsin) in the stomach at pH2.0.

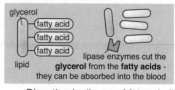

Digestion by lipase of fat and oil in the small intestine at pH7.0.

- Cells in the stomach produce hydrochloric acid which kills bacteria in food, unravels proteins to allow their digestion, supplies optimal acid conditions for the stomach's digestive enzymes, and activates pepsin (inactive before food arrives).

- **Bile** is a strong alkali produced by the **liver** and stored in the **gall bladder**. Bile joins partly digested food arriving in the small intestine. It neutralises the acid and provides the alkaline conditions the small intestinal enzymes need, and **emulsifies** (breaks down) fats.

Questions

Grades G-E
1 What is an 'active site'?

Grades D-C
2 Explain why enzymes do not work at extreme pHs.

Grades G-E
3 How does amylase help in digestion?

Grades D-C
4 How does bile help in digestion?

Enzymes at home

Biological washing powders

- Protein and fat stains on clothes can be difficult to wash out, requiring nearly boiling water. This can damage the clothes.

- Some microorganisms secrete protein- and fat-digesting enzymes from their cells. These enzymes are purified and added to **biological detergents** (washing powders). They work best at about 35–40 °C.
  - **Protease** enzymes digest protein (e.g. in blood and egg).
  - **Lipase** enzymes digest fat (e.g. butter and oil).

- Insoluble stains are difficult to remove from fabrics. In biological washing powders, **detergent** molecules make greasy stains **soluble**, lifting them from the fabric, and **enzymes** break them down.

- **Proteases** digest protein in stains.

- **Lipases** digest fat droplets when detergent frees them from fabric fibres and they pass into the water.

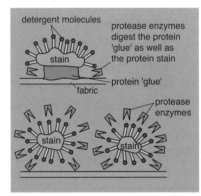

Removing a protein stain.

Enzymes and industry

Enzymes and the food industry

- Small babies do not digest proteins easily. Manufacturers use **proteases** to pre-digest the protein in baby foods.

- **Carbohydrases** are used to convert **starch** from potatoes and maize into **sugar syrups** (saturated sugar solutions) that are used in sports drinks.

- To reduce the sugar content of sugars in slimming foods, the enzyme **isomerase** is used to convert **glucose** into the sugar **fructose**, which tastes much sweeter.

How Science Works

You should be able to: evaluate the advantages and disadvantages of using enzymes in the home and in industry.

These are made with the help of enzymes.

- Small babies cannot break up protein molecules in their food into amino acids. **Baby food** manufacturers add **proteases** to do this. Babies can then absorb the amino acids into their bloodstream for transport to all their body tissues. Older babies produce all the digestive enzymes they need.

Questions

(Grades G-E)

1 How do proteases in washing powders help to remove blood stains?

(Grades D-C)

2 Explain how detergents make it easier for enzymes in washing powders to work.

(Grades G-E)

3 Why is fructose used in 'slimming' foods?

(Grades D-C)

4 Why are proteases sometimes used in the production of baby foods?

Respiration and energy

Energy: getting it and using it

- Energy for cells is supplied by **aerobic respiration**, a process that happens in cell organelles called **mitochondria**. The equation that summarises aerobic respiration is:

glucose + oxygen → carbon dioxide + water (+ energy)

- The energy is used:
  - to build large molecules from small ones
  - by plants to build sugars, nitrates and other nutrients into amino acids, then proteins
  - by mammals and birds to maintain a steady body temperature
  - by animals to make muscles **contract**.

- Energy from respiration allows:
  - muscles to contract, enabling movement
  - mammals and birds to maintain their body temperature in a colder environment
  - all living things to make larger molecules from smaller ones. In animals, this includes glycogen (storage carbohydrate) and proteins – see diagram.

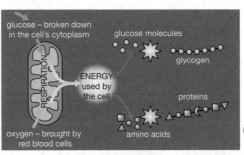

Small molecules join to make big ones.

Top Tip!

Do not confuse respiration and breathing. Respiration happens inside cells. Breathing is movement of air in and out of the lungs.

Removing waste: lungs

The lungs and gas exchange

- **Carbon dioxide** is a waste product of **respiration** in cells and must be removed. It **diffuses** from cells into the blood, and then diffuses from blood into the air sacs (alveoli) of our lungs.

- Breathing out (**expiration**) causes a pressure that forces the air, high in carbon dioxide, out of the lungs. Breathing in (**inspiration**) lowers the pressure and draws air into the lungs.

- Muscles for breathing include **intercostal muscles** between the ribs, which raise and lower them, and the **diaphragm**, which flattens as it contracts.

- To maximise carbon dioxide **diffusion** from bloodstream to lung cavity, **alveoli** have a rich blood supply, thin walls and a very large total surface area.

- In the opposite direction, **oxygen** from incoming air diffuses through alveolar walls into the bloodstream.

- The gases diffuse towards a region of lower concentration – they diffuse down a **concentration gradient**.

- The brain detects excess carbon dioxide in the blood. It sends nerve impulses to the intercostal and diaphragm muscles to make us breathe faster and more deeply.

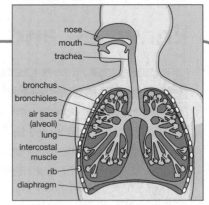

Structure of the lungs and muscles of breathing.

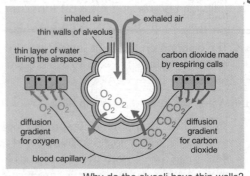

Why do the alveoli have thin walls?

Questions

(Grades G-E)

1 Without checking back, write the word equation for aerobic respiration.

(Grades D-C)

2 Why do mammals and birds need more energy than fish?

(Grades G-E)

3 Name the **two** sets of muscles involved in breathing.

(Grades D-C)

4 List **three** ways in which alveoli are adapted for quick gas exchange.

Removing waste: liver and kidneys

Producing urea and excreting urine

G–E

- During digestion, **protein** in the diet is broken down into **amino acids**. Cells then use these to build their own proteins.

- The **liver** converts any surplus amino acids into **urea**, a soluble substance which the blood transports to the **kidneys**.

- The kidneys filter waste substances from the blood, including urea, producing **urine** which flows to the **bladder** before being excreted.

Top Tip!

Remember: *urea* is made in the liver, not in the kidneys. The kidneys make *urine*, which contains urea dissolved in water.

How our bodies remove urea.

D–C

- The body's largest organ, the **liver**, breaks down many surplus and toxic substances. It makes urea, which then travels to the kidneys.

- In the kidneys, tiny tubes called **nephrons** collect waste substances from the blood, including urea and water. The waste (urine) flows down the ureters and into the bladder.

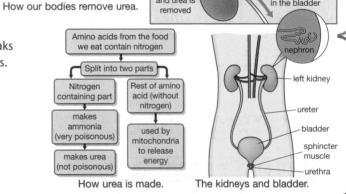

How urea is made.

The kidneys and bladder.

Homeostasis

Blood sugar, temperature and water

G–E

- To remain healthy, the body's **internal conditions** must be kept constant. This is called **homeostasis**.

- **Insulin**, secreted by the pancreas, enables cells to take in **blood sugar**, glucose, for respiration. Excess blood sugar can be fatal, and insulin also causes it to convert into glycogen.

- Enzymes in cells need a **temperature** of about 37 °C to work properly. When the body overheats, **sweating** cools it. When it is too cold, blood is re-routed deeper in the body.

- Cells gain **water** and **mineral ions** from food. Excess water could burst cells, and excess ions could be toxic, so the body regulates their concentration.

How homeostasis works.

D–C

- The diagram on the right shows how the body balances its **water** gains and losses to avoid cell dehydration or damage by excess water.

- Water loss in urine differs in hot and cold weather, as the table shows.

Water in, water out – it must balance.

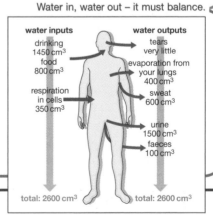

Summer and winter mean daily urine output.

day	1	2	3	4	5
mean summer output (cm³)	995	1000	975	1005	990
mean winter output (cm³)	1325	1295	1310	1300	1305

Questions

(Grades G-E)

1 Which organ changes surplus amino acids into urea?

(Grades D-C)

2 What is a nephron?

(Grades G-E)

3 List **three** internal conditions that are kept constant.

(Grades D-C)

4 Why is more urine often produced on a cold day than on a hot day?

Keeping warm, staying cool

Keeping close to 37 °C

- Normal **core body temperature** is between 35.5 and 37.5 °C.

- The **thermoregulatory centre** is in the brain. It contains **temperature receptors** that monitor blood temperature.

- Also, temperature receptors in the **skin** send nerve impulses to the thermoregulatory centre.

- If the blood is too hot or too cold, the centre sends impulses to parts of the body to make them restore the temperature to normal.

- If a person gets much too cold, the thermoregulatory centre stops working and **hypothermia** results: they cannot warm up. At 25 °C, the person will die.

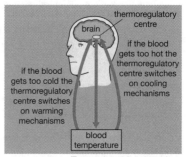

Temperature control is a homeostatic mechanism.

G–E

D–C

Treating diabetes

Glucose balance and insulin

- The **pancreas** monitors blood **glucose** and controls its level by producing the **hormone insulin**.

- Insulin allows cells, particularly muscle and liver cells, to absorb glucose and convert it to the storage material **glycogen**.

- It is life-threatening for blood glucose concentration to rise (or fall) too far.

- People with **diabetes** produce too little or no insulin. They may have to inject insulin regularly and follow a diet to control sugar intake.

- Diabetes symptoms include: frequent urinating and glucose in the urine, thirst, tiredness, weight loss and blurred vision.

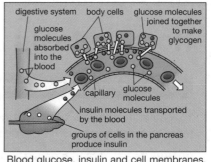

Blood glucose, insulin and cell membranes.

G–E

The history of controlling diabetes

- Scientists linked the islets of Langerhans, groups of cells in the pancreas, to diabetes.

- In 1921, Canadians Banting and Best collected fluid from the islets of Langerhans in the pancreas of a healthy dog. They injected it into a dog with diabetes, which was restored to health.

- The effective substance was named **insulin**, and since then has kept countless people with diabetes alive.

How Science Works

You should be able to:

- evaluate the data from the experiments by Banting and Best which led to the discovery of insulin

- evaluate modern methods of treating diabetes.

D–C

Questions

Grades G-E

1 Where is the body's thermoregulatory centre?

Grades D-C

2 Explain why a person with hypothermia cannot get warm.

Grades G-E

3 Which organ secretes the hormone insulin?

Grades D-C

4 How did we discover what insulin does?

Cell division – mitosis

Making new body cells

G–E

- Cells divide in two to produce more cells, so that the body grows and replaces worn and dead cells.

- A human **body cell** has two sets of 23 **chromosomes**, one set from each parent, therefore 23 pairs totalling 46. (Sex cells, gametes, have only one set.) Chromosomes contain **genes** that control inherited characteristics.

- In cell division, chromosomes must be duplicated so that each new cell contains the identical two sets of chromosomes.

Stages in mitosis

D–C

- The process of making new body cells is **mitosis**, shown in the diagram.

- Some organisms reproduce **asexually**. They produce new cells by mitosis. The new cells grow into a new organism. It is genetically identical to its parent.

Top Tip!

Cells produced by mitosis have identical genes.

membrane round nucleus

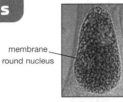
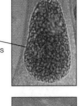

chromosomes in the nucleus

copies of chromosomes

1 Cell before mitosis starts

2 The chromosomes make exact copies of themselves

chromosomes move apart

4 The copies of the chromosomes move apart

3 The nuclear membrane breaks apart

the chromosomes are at opposite ends of the cell

new nucleus in each cell

5 The cell starts to split in two

6 Two identical cells, they contain the same genes

Gametes and fertilisation

How gametes are formed

G–E

- **Gametes** are formed in the reproductive organs (**ovaries** and **testes**).

- A gamete must have only *half* the body cell number of chromosomes so that, when two gametes fuse in **fertilisation**, the cell formed contains the normal number.

- In humans, the new cell formed at fertilisation contains two sets of chromosomes, 23 from the sperm and 23 from the egg. The cell divides over and over again by mitosis, to form a new organism.

Importance of fertilisation

D–C

- Genes determine inherited characteristics. **Alleles** are different varieties of a gene. A person has two copies of each gene, one from each parent. Different alleles determine different variations of a particular characteristic.

- The alleles in sperm and egg are brought together at fertilisation. So an offspring inherits alleles, and hence characteristics, from both parents. The mixture of alleles is different from the mixture in either of the parents, so each offspring has a unique set of genes. This is one cause of **variation**.

Questions

(Grades G-E)
1 How many sets of chromosomes are there in a body cell?

(Grades D-C)
2 Name the kind of cell division that makes new cells for growth, repair and asexual reproduction.

(Grades G-E)
3 Why must sperm and eggs contain only one set of chromosomes each?

(Grades D-C)
4 Why are offspring produced by sexual reproduction not exactly like their parents?

Stem cells

Unspecialised, specialised and stem cells

- Some cells start by being **unspecialised**: they could develop into any type of cell. Then in **animals** they **differentiate** to do just one job – they become **specialised**.

- **Stem cells**, in embryos and certain parts of adults, are undifferentiated. They can divide by mitosis and develop into different cell types.

- Scientists hope to use stem cells to repair damaged body parts (e.g. nervous tissue).

- Many **plant** cells can differentiate into any type of cell throughout their life.

Adult and embryo stem cells

- Different stem cells in **bone marrow** can develop:
  - either into red or white blood cells
  - or into bone cells, cartilage cells, tendon cells or fat cells.

- Stem cells from 4- to 5-day-old **embryos** placed in **cell growth medium** can divide for months without differentiating. Treatment with particular proteins turns some genes on or off, and the cells then differentiate into different tissues.

How Science Works

You should be able to: make informed judgements about the social and ethical issues concerning the use of stem cells from embryos in medical research and treatments.

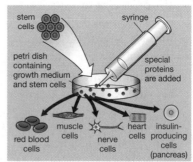

Embryo stem cells can become many types of cells.

Chromosomes, genes and DNA

Chromosomes, DNA and genes

- A **chromosome** is a very long molecule of **deoxyribonucleic acid** (**DNA**). The structure of DNA was discovered by Francis Crick and James Watson.

- There are 46 chromosomes in the **nucleus** of each human body cell.

- A gene is a tiny section of a chromosome. The genes contain the **genetic code**, the instructions for joining amino acids together to make proteins.

- Many proteins are **enzymes**. Thousands of enzymes control the chemical reactions of all the cells in our body, and thereby determine our characteristics.

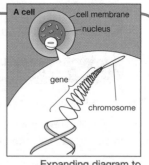

Expanding diagram to illustrate DNA.

How DNA copies itself in mitosis

- The diagram shows part of the DNA double helix with its complementary bases.

- Before **mitosis** begins (see page 85), the base pairs separate, unzipping the two strands. Complementary bases move to each single strand and attach themselves, making two double strands.

- The strands coil up to form two identical chromosomes, ready to separate into two new cells.

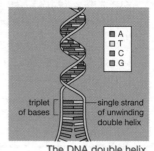

The DNA double helix.

Questions

1 What is a stem cell?

2 Name **two** different kinds of cells that stem cells in bone marrow can produce.

3 What is the chemical that chromosomes are made of?

4 Why must DNA duplicate itself before mitosis begins?

Inheritance

Mendel's experiments on breeding pea plants

How Science Works

You should be able to:
• explain why Mendel proposed the idea of separately inherited factors and why the importance of this discovery was not recognised until after his death
• interpret genetic diagrams.

• Mendel repeated same-colour crosses of pea plants until the offspring had either red only flowers or white only flowers. These were **pure breeding** plants, and Mendel concluded that a 'factor' from each parent determined flower colour.

• The 'factor' is a **gene**. The **characteristic** of flower colour is controlled by a *single gene* which has two forms or **alleles** – red and white. A plant inherits one allele from each parent. Where both alleles are the same, the plant is **homozygous**. Where they are different, it is **heterozygous**. These rules apply to all characteristics controlled by a single gene with two alleles.

• Mendel then crossed homozygous red plants with homozygous white plants and all the offspring had red flowers.

This is a Punnett square. It shows how the genes in this cross are inherited.

Parents:	Rr		Rr
Gametes:	ⓡ and ⓡ		ⓡ and ⓡ
Offspring:		ⓡ	ⓡ

	ⓡ	ⓡ
ⓡ	**RR** red flowers	**Rr** red flowers
ⓡ	**Rr** red flowers	**rr** white flowers

• All these offspring inherited a white and a red allele. But red was **dominant**. White was **recessive** and would show only if red was absent.

• Mendel then crossed these red offspring with each other. Three-quarters of the offspring had red flowers and one-quarter had white flowers. He concluded that a characteristic may not show but can still pass to the next generation. The genetic diagram on the right shows what happened. **R** is the dominant allele for red flowers and **r** is the recessive allele for white flowers.

Top Tip!

Genetic diagrams do not tell you how many offspring two parents will have. They just tell you the chances of getting different combinations of alleles.

How is sex inherited?

Sex inheritance and sex ratio

• Human **body cells** contain 23 chromosome pairs. The pair of **sex chromosomes**, **X** and **Y**, carry the genes which determine sex.

• An egg or sperm contains one sex chromosome only.

• As the table shows, an embryo always inherits X from the egg. If the sperm supplies: an X, it will be a girl (XX); a Y, it will be a boy (XY).

How sex is determined.

sex chromosome		sex chromosomes	
in egg	in sperm	in embryo	sex of embryo
X	X	XX	female
X	Y	XY	male

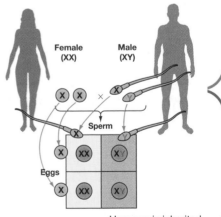

How sex is inherited.

• As the diagram shows, a sperm has an equal chance of containing the X or the Y chromosome. The sperm determines a child's sex, and all eggs have an equal chance of fusing with either X or Y.

• The same number of males as females would be expected. This is roughly the case.

Questions

(Grades D-C)
1 What does 'pure-breeding' mean?

(Grades D-C)
2 Explain what is meant by a 'recessive allele'.

(Grades G-E)
3 Which determines the sex of a baby – the sperm or the egg?

(Grades D-C)
4 Why are roughly equal numbers of boy babies and girl babies born?

Inherited disorders

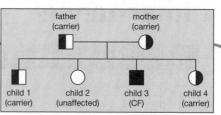

Inheritance of cystic fibrosis.

Inherited disorders

- An **inherited disorder** prevents the body from working properly. The disorder lasts for life because chromosomes in all body cells carry the gene.

- A **recessive allele** of a certain gene causes **cystic fibrosis**, which affects cell membranes. Two healthy parents may be **carriers** and may each pass on a recessive allele to their child.

- A **dominant allele** causes **Huntington's disease** which affects the nervous system. A child will inherit Huntington's disease if it gets this allele from just one parent.

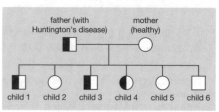

Inheritance of Huntington's disease.

Cystic fibrosis and its treatment

- A child with cystic fibrosis (CF) but who has healthy parents inherits the recessive allele from both parents. They are protected by the dominant (healthy) allele, but half their gametes will carry the recessive allele.

- A person with CF cannot digest food properly, has frequent chest infections, fertility problems and a shortened life.

- There is no cure, but treatments can minimise the symptoms and tackle infections.

- Embryos can be **screened** to check for this disorder.

How Science Works

You should be able to: make informed judgements about the economic, social and ethical issues concerning embryo screening that you have studied, or from information you have received.

DNA fingerprinting

What is a DNA fingerprint?

- All your body cells contain the same **DNA**. It differs from anyone else's except an identical twin – it is **unique** and cannot be changed.

- DNA (from e.g. saliva, hair, blood, semen) displayed on a sheet of film is called a **DNA** (or **genetic**) **fingerprint** or a **DNA profile**. It:
  - requires only a tiny sample
  - identifies one person only, alive or dead
  - can be used to identify close relations.

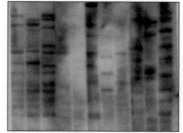

A good DNA fingerprint identifies just one person.

Making and using DNA fingerprints

- In **DNA fingerprinting**, DNA is cut into different lengths by enzymes. The bits are separated into bands using a technique like chromatography. The bands are treated with radioactive chemicals to make an image of them on X-ray film.

- In **solving crime**, DNA from the crime scene can be matched to DNA from a suspect.

- In **paternity disputes**, the number of bands shared by a man and a child shows whether he is the child's father.

Questions

(Grades G-E)

1 Give **two** examples of inherited disorders.

(Grades D-C)

2 How can two parents without CF have a child with CF?

(Grades G-E)

3 Why can a DNA fingerprint identify a single person?

(Grades D-C)

4 Name **two** uses of DNA fingerprinting.

B2b summary

Enzymes are **biological catalysts** that speed up chemical reactions and are not used up.

Enzymes are involved in **respiration, photosynthesis** and **protein synthesis**.

Each enzyme works at an **optimum pH** and **temperature**.

Enzymes

The digestive enzymes **amylases**, **lipases** and **proteases** break down food.

Enzymes are used in industry in **food production** and in the home in **biological detergents**.

Aerobic respiration is carried out in **mitochondria** in our cells to provide energy:
glucose + oxygen → carbon dioxide + water + ENERGY

Oxygen for respiration **diffuses** into the blood from the lungs. Carbon dioxide from respiration diffuses in the opposite direction, and is removed from the lungs when we breathe out. The dissolved gases are carried round the body in the blood.

Respiration and waste removal

Waste proteins are broken down by the **liver** into **urea**.

The **kidneys** make **urine**, which contains dissolved urea.
The urine is stored in the **bladder** then excreted.

Blood sugar levels are controlled by **insulin** produced by the **pancreas**.
People with **diabetes** produce too little insulin or do not respond to it.

Human body **temperature** is kept constant at about 37 °C.
The **thermoregulatory** centre in the brain contains **temperature receptors**.

Homeostasis

Sweating cools the body down.
Rerouting blood deeper in the body reduces heat loss.

Water inputs from drinking and cell respiration are balanced with water outputs from tears, evaporation from the lungs, sweat, urine and faeces.

Chromosomes are long molecules of **DNA** made up of **genes**.
Genes contain the **genetic code**, which contains the instructions for making proteins, including **enzymes**.

Body cells divide by **mitosis**. Each new cell has the same number of identical chromosomes as the original cell.

Inheritance

Some inherited **characteristics** are controlled by a single gene.
Different forms of a gene are called **alleles**.

In **homozygous** individuals, the alleles inherited from each parent are the same.
In **heterozygous** individuals, they are different.

Alleles can be **recessive** or **dominant**.

Gametes (sperm and egg) fuse at **fertilisation** to produce a new organism.
The mixture of **alleles** from the parents causes **variation** in the offspring.

Undifferentiated **stem cells** can **specialise** into many types of cells.

Sex chromosomes determine the sex of the offspring (male XY, female XX).

Atomic structure 2

The structure of an atom

- An **atom** contains **sub-atomic particles** (see also page 26). Its small central **nucleus** contains **protons** and **neutrons**. Round the nucleus are **electrons**.
  - For the **relative masses** and **charges** of sub-atomic particles, see the table on page 26.
- In an atom, the number of electrons = the number of protons. It has no overall electrical charge.
- All atoms of an **element** have the same number of protons. Atoms of different elements have different numbers of protons.
- For information on the **symbol**, **mass number** and **atomic number** of an element, see page 26.
- The elements are arranged in the periodic table (see page 26) in order of atomic number.

G–E

D–C

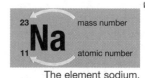

The element sodium.

Electronic structure

The arrangement of electrons

- Electrons are arranged round a nucleus in **shells**. Each shell represents a different **energy level**. The shells closest to the nucleus have the lowest energy levels.
- An electron always takes up the lowest available energy level, in the innermost shell available.
- The first shell holds up to two electrons. The second shell holds up to eight electrons.
- Atoms with a full outer electron shell are stable and unreactive.
- When an atom reacts, it gains, loses or shares one or more electrons to achieve a full outer electron shell.

one electron in the third shell
two electrons in the first shell

Na

eight electrons in the second shell

Arrangement of electrons in a sodium atom. Sodium reacts by losing **one** electron and so is in Group **1** of the periodic table.

G–E

How Science Works

You should be able to: represent the electronic structure of the first 20 elements of the periodic table in the following forms.

For sodium: and 2, 8, 1

Electron notation

- A **magnesium** atom has 12 electrons.
- Shells 1 and 2 contain the maximum two and eight electrons. The remaining two electrons occupy the third shell.
- The **electron notation** for a magnesium atom is 2, 8, 2.
- Magnesium reacts by losing the **two** electrons in its outermost shell, and so is in Group **2** of the periodic table.
- A **calcium** atom has 20 electrons. Its electron notation is 2, 8, 8, 2.
- Calcium reacts by losing its outermost **two** electrons, placing it in Group **2**.

two electrons in the third shell
two electrons in the first shell

Mg

eight electrons in the second shell

Arrangement of electrons in a magnesium atom.

D–C

Arrangement of electrons in a calcium atom.

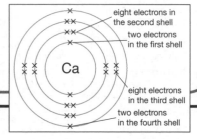

eight electrons in the second shell
two electrons in the first shell

Ca

eight electrons in the third shell
two electrons in the fourth shell

Questions

1 State the charges on each of these sub-atomic particles: a proton, a neutron, an electron.

2 Look at the periodic table on page 26. What is the mass number of potassium? What does this tell you about the structure of a potassium atom?

3 A potassium atom has 19 electrons. The inner shell contains two electrons, the next shell contains eight and the next shell also contains eight. How many electrons are there in the outer shell of a potassium atom?

4 How can a potassium atom become stable?

Mass number and isotopes

Elements, groups and relative atomic mass

- Atoms of an **element** all have the same number of **protons** and **electrons**. An atom has no net charge. **Isotopes** of an element have different numbers of neutrons (see also page 127).

- The **periodic table** arranges the elements in order of **atomic number**.

- A **group** contains elements with the same number of electrons in their outermost shell (the highest energy level). This determines the reactions an element can undergo.

- The **mass number** of an element is the sum of protons and neutrons.

- For most elements, the mass number is same as the element's **relative atomic mass** or A_r for short.

Ionic bonding

Ion formation and ionic compounds

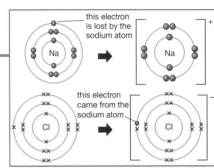

When a sodium atom becomes a sodium ion, its electron arrangement changes from 2, 8, 1 to $[2, 8]^+$. When a chlorine atom becomes a chloride ion, its electron arrangement changes from 2, 8, 7 to $[2, 8, 8]^-$.

- To form **ionic compounds**, the atoms of reacting elements either lose electrons and become **positively** charged **ions**, or gain electrons and become **negatively** charged ions.

- An **ionic bond** is the force of attraction between oppositely charged ions. Ionic bonds hold ionic compounds together strongly, making them solids at room temperature.

- All ions achieve a full outermost electron shell, giving the electronic structure of a noble gas (Group 0).

- **Metal** atoms *lose* electrons to form positive ions.

- **Non-metal** atoms *gain* electrons to form negative ions.

- Diagram 1 shows the formation of ions when magnesium metal reacts with oxygen gas to form the solid, magnesium oxide.

How Science Works

You should be able to: represent the electronic structure of the ions in sodium chloride, magnesium oxide and calcium chloride in the following forms.

For sodium ion (Na⁺): and $[2, 8]^+$

- Diagram 2 shows the formation of ions when calcium metal is heated in chlorine gas to form the solid, calcium chloride.

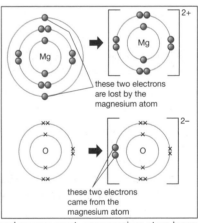

these two electrons are lost by the magnesium atom

these two electrons came from the magnesium atom

1 When magnesium and oxygen react, a magnesium atom loses two electrons to form an ion with a 2+ charge, and an oxygen atom gains two electrons to form an ion with a 2– charge.

2 When calcium and chlorine react, a calcium atom loses two electrons to form an ion with a 2+ charge, and *each of two* chlorine atoms gains one electron to form an ion with a 1– charge.

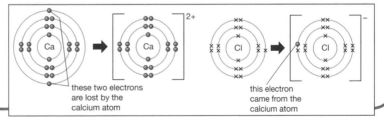

these two electrons are lost by the calcium atom

this electron came from the calcium atom

Questions

1 A lithium atom has three protons. How many electrons does it have? What is its charge?

2 What do all the elements in the same group in the periodic table have in common?

3 A potassium atom has one electron in its outer shell. A chlorine atom has seven electrons in its outer shell. How can each of these atoms become stable?

4 Explain why the *overall* charge of the ions in an ionic compound is zero.

Ionic compounds

The structure and properties of ionic compounds

- An **ionic compound** contains **positive metal ions** and **negative non-metal ions**.

- Electrostatic **forces of attraction** form strong bonds between oppositely charged ions. (See also page 27.)

- Ionic compounds:
  - are solids at normal temperatures
  - do not conduct electricity as solids
  - have high melting points and boiling points.
  - conduct electricity when molten
  - conduct electricity when dissolved (in water)

- Ionic compounds are **giant ionic structures**.

- At normal temperatures the strong electrostatic forces of attraction act in all directions, fixing the positions of oppositely charged ions in an ordered **lattice**.

- An electric current is the movement of charged particles. Ionic compounds can **conduct electricity** when **molten** because positive and negative ions are then free to move independently.

- Many ionic compounds are soluble in water. Then, the ions can move independently. This enables the **solution** to conduct electricity.

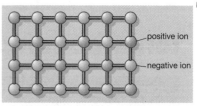

positive ion
negative ion

Lattice structure of an ionic compound. Ionic bonds fix each ion in position next to oppositely charged ions only.

How Science Works

You should be able to: suggest the type of structure of a substance, given its properties.

Covalent bonding

Non-metal atoms in compounds

- A **covalent bond** is formed when two atoms **share** electrons. Covalent bonds are very strong.

- Covalent bonds hold atoms together in **simple molecules** such as hydrogen, chlorine, oxygen, hydrogen chloride, water and ammonia.

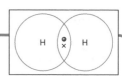

The atoms in a hydrogen molecule are held together by a covalent bond. The molecule is represented as H_2 or H–H.

Covalently bonded simple molecules

- The first electron shell is **stable** when it contains two electrons, and the second when it contains eight electrons.

- Atoms in covalent compounds share electrons to achieve stable outermost shells of electrons.

An oxygen atom has 6 electrons in its outermost shell. To make 8, each oxygen atom shares 2 electrons with the other oxygen atom, making a **double** bond.

A molecule of oxygen, O_2, O=O.

An oxygen atom shares 1 electron from each hydrogen atom to make 8, and each hydrogen atom shares 1 electron from oxygen to make 2.

A molecule of water, H_2O, H–O–H.

How Science Works

You should be able to: represent the covalent bonds in molecules such as water, ammonia, hydrogen, hydrogen chloride, chlorine, methane and oxygen.

For ammonia (NH_3):

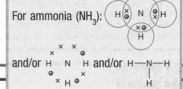

and/or

Questions

Grades G-E
1 Name **two** states in which an ionic compound can conduct electricity.

Grades D-C
2 Explain what is meant by a 'giant ionic structure'.

Grades G-E
3 Name the kind of bond that is formed when two atoms share a pair of electrons.

Grades D-C
4 Explain why two oxygen atoms can both become stable if they share two pairs of electrons.

Simple molecules

Simple molecules and their properties

- **Simple molecules** contain a few atoms joined by strong **covalent bonds** (see page 92).

- **Elements** include:
  - hydrogen, H_2, H—H
  - oxygen, O_2, O=O
  - chlorine, Cl_2, Cl—Cl

- **Compounds** include:
  - water, H_2O, H—O—H
  - hydrogen chloride, HCl, H—Cl
  - ammonia, NH_3, H—N—H with H below
  - methane, CH_4, H—C—H with H above and below

> **Top Tip!**
> Simple molecules consist of small numbers of atoms held together by shared pairs of electrons forming covalent bonds.

- **Properties** of simple molecular compounds:
  - they do not conduct electricity
  - they have low melting and boiling points
  - they are often liquids or gases at room temperature.
  - when solid, they are brittle and often soft or waxy
  - they are often insoluble in water but may dissolve in other liquids

- Covalently bonded simple molecules have no charged particles. Therefore they do not conduct electricity.

G–E

D–C

Giant covalent structures

Giant covalent structures and their properties

- **Giant covalent structures** are macromolecules. Strong covalent bonds join many atoms arranged in an orderly, three-dimensional **lattice**.

- These macromolecules have high melting points. Many are very hard. They do not conduct electricity or dissolve in water.

- Examples include the element carbon in the form of **diamond** and **graphite** and the compound **silicon dioxide**.

> **Top Tip!**
> In general, covalent bonds are just as strong as ionic bonds.

G–E

Diamond and silicon dioxide

> **How Science Works**
> You should be able to: represent the covalent bonds in giant structures, e.g. diamond and silicon dioxide.

- **Diamond** is a form of the **element** carbon and is the hardest **mineral**. Very strong covalent bonds join each carbon to four others, which is what makes diamond so hard.

A small part of the diamond lattice. All carbon atoms are strongly bonded to four others.

- **Graphite** is also a form of carbon. In each atom, three electrons form covalent bonds with other carbon atoms, giving a layered arrangement. The layers slide over each other easily, so graphite is soft and slippery.

Graphite has free electrons between layers.

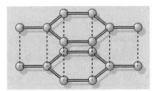

- The **compound silicon dioxide** is sand. It is very hard and has a high melting point. It resists weathering because its silicon-oxygen covalent bonds are very strong.

A small part of the silicon dioxide lattice.

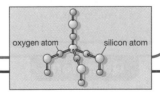

oxygen atom   silicon atom

D–C

Questions

(Grades G-E)
1 Name **two** elements that exist as simple molecules.

(Grades D-C)
2 Explain why compounds made of molecules do not conduct electricity.

(Grades G-E)
3 Give **one** example of an element that can form a giant covalent structure.

(Grades D-C)
4 Explain why graphite is soft.

Metals

How the structure of metals relates to their properties

- **Metals** have few electrons in their outer shells. They are lost when metals form **positive** ions.

- Metals can be bent and shaped because the layers of atoms easily slide over each other.

- Heavier metals have high densities and strength, and high melting and boiling points.

- The properties of metals determine the many uses that metals are put to.

- When a potential difference (voltage) is placed across some metal, the delocalised electrons move, carrying the electrical **charge** through the structure.

- Heating increases the kinetic energy of the delocalised electrons in the metal. They move faster, transferring **thermal energy** to positive ions. In this way, heat moves rapidly through the metal by conduction.

An atom of the metal element magnesium. It loses the two outer electrons when it reacts.

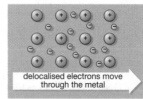

delocalised electrons move through the metal

In a circuit, delocalised electrons move in one direction to carry charge to the positive terminal of a cell.

G–E

D–C

Alkali metals

Group 1 – the alkali metals

- The elements of **Group 1** are the **alkali metals**.

- Alkali metals are very reactive. All react vigorously with water. Reactivity increases down the group.

- In reactions, they lose one outer electron to form **ions** with a 1+ charge.

- They react with non-metals to form **salts** which are **ionic compounds**.

Group								$^1_1$ H hydrogen
I	II							
$^7_3$ Li lithium	$^9_4$ Be beryllium							
$^{23}_{11}$ Na sodium	$^{24}_{12}$ Mg magnesium							
$^{39}_{19}$ K potassium	$^{40}_{20}$ Ca calcium	$^{45}_{21}$ Sc scandium	$^{48}_{22}$ Ti titanium	$^{51}_{23}$ V vanadium	$^{52}_{24}$ Cr chromium	$^{55}_{25}$ M mangan		
$^{85}_{37}$ Rb rubidium	$^{88}_{38}$ Sr strontium	$^{89}_{39}$ Y yttrium	$^{91}_{40}$ Zr zirconium	$^{93}_{41}$ Nb niobium	$^{96}_{42}$ Mo molybdenum	$^{99}_{43}$ T technetii		
$^{133}_{55}$ Cs caesium	$^{137}_{56}$ Ba barium	$^{139}_{57}$ La lanthanum	$^{178}_{72}$ Hf hafnium	$^{181}_{73}$ Ta tantalum	$^{184}_{74}$ W tungsten	$^{186}_{75}$ R rheniu		
$^{223}_{87}$ Fr francium	$^{226}_{88}$ Ra radium	$^{227}_{89}$ Ac actinium						

The Group 1 metals are on the far left of the periodic table.

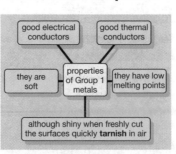

Properties of Group 1 metals.

How alkali metals react down Group 1

- When an alkali metal loses one electron, the ion is left with the full outer shell of a noble gas.

A lithium ion has the configuration of the noble gas helium and a charge of 1+.

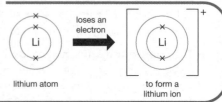

lithium atom

to form a lithium ion

D–C

G–E

Questions

Grades G-E
1 How do metals form ions?

Grades D-C
2 Explain why metals can be bent and shaped.

Grades G-E
3 How many electrons are there in the outer shell of the Group 1 metals?

Grades D-C
4 Which is more reactive – sodium or rubidium? Explain your answer.

Halogens

Group 7 – the halogens

- The elements of Group 7 are the halogens:
  - **fluorine**, pale yellow gas
  - **chlorine**, green gas
  - **bromine**, dark red liquid
  - **iodine**, dark grey crystalline solid
  - **astatine**, highly radioactive solid.

- All exist as molecules with two covalently bonded atoms. (See also page 92.)

- On heating, solid **iodine** sublimes to give a purple vapour.

- With alkali metals, halogens form **salts** which are **ionic compounds**.

- Halogens have seven electrons in their outermost shell. They react by gaining one electron to acquire the electron structure of a noble gas.

- Sodium metal burnt in chlorine gas forms the **ionic compound** sodium chloride: **sodium + chlorine → sodium chloride**

- Halogen ions are called **halide** ions. They have a 1– charge.

- Fluorine is the halogen with the smallest atoms and is the most reactive. Reactivity decreases *down* Group 7.

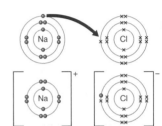

When chlorine atoms react they gain an electron to form a chloride ion which has a 1– charge and the electron structure of noble gas argon.

Nanoparticles

What is a nanoparticle?

- A **nanoparticle** is a tiny manufactured structure of a few hundred atoms, from 1 to about 100 nanometres long. (A human hair is 100 000 nanometres wide.)

- The atoms in nanoparticles are regularly arranged in hollow structures such as tubes and spheres one atom thick.

- Being so small, nanoparticles have a very high **surface area to volume** ratio.

- Nanoparticles are likely to find uses in computers, as catalysts, sensors and coatings, and as new materials.

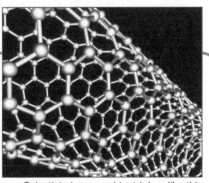

Scientists have used test tubes like this nanoparticle to catalyse new reactions.

Nanoscience using nanoparticles

- Scientists expect to make **materials** from nanoparticles that are harder, lighter and stronger than any known materials.

- Nanoparticles may make good **biosensors**, detecting very low levels of chemicals and biological agents.

- Nanoparticles could be designed to **process information** in computers millions of times faster than current components.

- Nanoparticle **catalysts** are expected to speed up manufacturing processes.

Questions

(Grades G-E)
1 What are the halogens?

(Grades D-C)
2 How do the halogens form ions?

(Grades G-E)
3 Explain what is meant by 'nanoparticles'.

(Grades D-C)
4 How might nanoparticles be used to speed up chemical reactions?

Smart materials

Smart materials

- A **smart material** has a particular **property**: when there is a change in the **environment**, it **responds** by changing in some way that is **reversible**.

- Smart materials include alloys that have shape memory (see page 33), photochromic materials, thermochromic materials and electroluminescent materials.

G–E

Types and uses of smart materials

- **Photochromic** materials darken on exposure to strong light and are used in spectacles and sunglasses.

- **Thermochromic** materials respond to changes in **temperature** by changing **colour**. These materials are black at most temperatures but change colour over a particular range of temperatures. They are used in paper thermometers for use on children's foreheads.

- **Electroluminescent** materials emit light of different colours when an alternating current passes through them. These materials form a layer on the plastic sheeting of some advertising signs.

How Science Works

You should be able to:
- relate the properties of substances to their uses
- evaluate the developments and applications of new materials, e.g. nanomaterials and smart materials.

D–C

Compounds

Compounds and ratios of elements in them

- A **compound** contains atoms of more than one **element**.

- Atoms of two or more elements react to form a **compound**. The elements are **chemically combined** and are joined by **chemical bonds**.

- **Glucose** has a large **covalently bonded** molecule with the **chemical formula** $C_6H_{12}O_6$. The letters indicate the elements in glucose, and the numbers show how many atoms of each element there are. The **ratio** of elements is: $6:12:6 = 1:2:1$.

- Salt, **sodium chloride**, has a **giant ionic structure** of formula NaCl. Its sodium ions (Na^+) and chloride ions (Cl^-) are in the **ratio** 1:1.

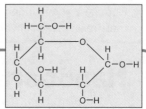

This shows the way that atoms are arranged in one molecule of glucose. The lines between the atoms represent chemical bonds.

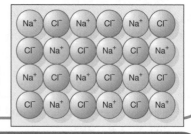

Each grain of salt contains a huge number of sodium and chloride ions.

G–E

Mixtures and compounds

- A **mixture** is composed of different substances that are *not* chemically combined. The substances can be mixed in any proportion. They are not in fixed ratios. They can be separated.

- In a **compound**, elements are chemically joined in particular, fixed ratios.

In water, hydrogen and oxygen are chemically combined in the ratio of 2:1.

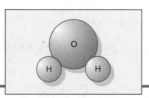

Substances in this mixture are not chemically combined or in fixed ratios.

D–C

Questions

(Grades G-E)
1 Explain what is meant by a 'smart material'.

(Grades D-C)
2 Explain how a photochromic material behaves when exposed to light.

(Grades G-E)
3 The chemical formula of methane is CH_4. What does this tell you about the numbers and types of atoms in a methane molecule?

(Grades D-C)
4 Explain the difference between a 'compound' and a 'mixture'.

Percentage composition

Relative formula mass and percentage of elements in compounds

- Each element has a **relative atomic mass** (see page 91).
- To find the **relative formula mass** (M_r) of a molecule or compound, the relative atomic masses of all the atoms are added together in the **numbers** shown by the formula (see page 96).

  The formula of an oxygen **molecule** = O_2

  Relative atomic mass of an oxygen atom = 16

  Relative formula mass of $O_2 = 2 \times 16 = 32$

- The formula used to calculate the **percentage mass** of an **element** in a **compound** is:

$$\text{% mass of an element in a compound} = \frac{\text{relative atomic mass of the element} \times \text{number of atoms of element in the formula}}{\text{relative formula mass}} \times 100$$

- The formula of the **compound** potassium nitrate is KNO_3.

  Relative atomic masses: K = 39, N = 14, O = 16

  Relative formula mass of $KNO_3 = 39 + 14 + (3 \times 16) = 101$

  % mass of potassium in potassium nitrate $= \frac{39 \times 1}{101} \times 100 = 38.6\%$

How Science Works

You should be able to: calculate chemical quantities involving formula mass (M_r) and percentages of elements in compounds.

Moles

What is a mole?

- Atoms and molecules are so tiny that their masses cannot be measured individually. Instead, we use an amount called the mole:

  a mole of a substance is the relative formula mass in grams of that substance

  – There is the same number of particles in 1 mole of any substance.

- Formula of the compound water = H_2O

  Relative formula mass of $H_2O = (2 \times 1) + (1 \times 16) = 18$

  So the mass of 1 mole of water molecules is 18 g

The mass of 1 mole of water molecules is 18 g.

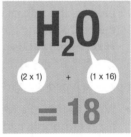

H_2O

(2 x 1) + (1 x 16)

= 18

- Water is formed when hydrogen is burnt in air (oxygen):

  hydrogen + oxygen → water

  The formula for hydrogen is H_2 and for oxygen it is O_2. Water is H_2O.

  The balanced equation is: $2H_2 + O_2 \rightarrow 2H_2O$

  This shows that 2 moles of hydrogen molecules react with 1 mole of oxygen molecules.

Top Tip!

The particles in a mole of a substance can be atoms or molecules (or ions).

- Magnesium oxide is formed when magnesium metal burns:

  magnesium + oxygen → magnesium oxide

  Magnesium oxide is MgO.

  The balanced equation for the reaction of magnesium and oxygen is: $2Mg + O_2 \rightarrow 2MgO$

  This shows that 1 mole of oxygen molecules reacts with 2 moles of magnesium atoms.

Top Tip!

The percentage composition is calculated by mass.

Questions

Grades G-E

1 The formula of methane is CH_4. The relative atomic mass of carbon is 12. The relative atomic mass of hydrogen is 1. Calculate the relative formula mass of methane.

Grades D-C

2 Calculate the per cent by mass of carbon in methane.

Grades G-E

3 What is the mass of 1 mole of methane? (Hint – look at your answer to question 1.)

Grades D-C

4 The equation for the reaction between carbon and oxygen is: $C + O_2 \rightarrow CO_2$. How many moles of carbon atoms react with each mole of oxygen molecules?

Yield of product

The amount of product obtained in a reaction

- No atoms are gained or lost in a chemical **reaction**. Yet the **reactants** that take part in the reaction do not always give as much **product** as expected. In other words, the **actual** amount can be less than the **theoretical** amount. The diagram shows some reasons for this.

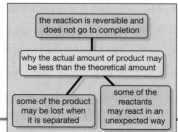

Why the actual amount of product in a reaction may be less than the theoretical amount.

Theoretical yield and actual yield

- The maximum amount of product from a reaction is the **theoretical yield**. The amount that actually forms is the **actual yield**.

- To work out the cost of making a chemical, a manufacturer first calculates how much product to expect from a reaction process.

- A reaction is said to have a high **atom economy** if a high proportion of the reactants end up as useful products.

- It is important for **sustainable development** and to limit costs to use reactions with a high atom economy.

Reversible reactions

Examples of reversible reactions

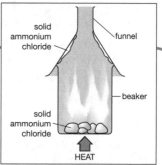

The reversible reaction of ammonium chloride.

- In some chemical reactions, the **products** can themselves react to produce the original **reactants**. Such reactions go **forwards** to give the products, and in **reverse** to give the reactants again. They are **reversible reactions**, represented as:

A + B ⇌ C + D

where A and B are reactants and C and D are products.

- When heated, the white solid **ammonium chloride** forms the gases **ammonia** and **hydrogen chloride**:

ammonium chloride ⇌ ammonia + hydrogen chloride

When the gases cool down, they react to re-form ammonium chloride.

> **Top Tip!**
> You need to be able to explain what a reversible reaction is.

- Copper sulfate reacts with water to give blue **hydrated copper sulfate**. When it is heated, the water is lost and it becomes white **anhydrous copper sulfate**. This reaction is reversible:

hydrated copper sulfate ⇌ anhydrous copper sulfate + water

- **Anhydrous cobalt chloride** is blue. It reacts with water to form pink **hydrated cobalt chloride**. This reaction is used to test whether water is present, and can be reversed by heating:

hydrated cobalt chloride ⇌ anhydrous cobalt chloride + water

Questions

(Grades G-E)
1 What can you say about the total number of atoms in the reactants and the products of a chemical reaction?

(Grades D-C)
2 State **two** reasons why the quantity of a product from a reaction may not be as great as expected.

(Grades G-E)
3 The breakdown of ammonium chloride to ammonia and hydrogen chloride is a reversible reaction. How can you show this when you write the equation for the reaction?

(Grades D-C)
4 Nitrogen and hydrogen can react to produce ammonia. This is a reversible reaction. Write the word equation for this reaction.

Grades

G–

D–

G–E

D–C

Equilibrium 1

Reversible reactions

• The **reversible reaction** between nitrogen and hydrogen forms ammonia:

nitrogen + hydrogen ⇌ ammonia

• A reversible reaction does not go to **completion**. The **forward reaction** and the **reverse** reaction take place at the same time.

The effects of reaction conditions on reversible reactions

• **Reaction conditions**, especially pressure and temperature, affect the yield of product in a reversible reaction.

• **Removing products** increases the forwards reaction.

• **Catalysts** increase the rate of a reversible reaction. They do *not* affect the amount of reactants and products, but will shorten the time taken for the reaction to reach a balance.

Haber process

The Haber process for making ammonia

• Ammonia is used to make fertilisers. Ammonia is produced by the **Haber process** in the reaction:

nitrogen + hydrogen ⇌ ammonia

• Nitrogen is obtained from the air and hydrogen comes from natural gas.

• The **conditions** used in the Haber process are:
  – an iron catalyst
  – a pressure of about 200 atmospheres
  – a temperature of about 450 °C.

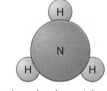

An ammonia molecule contains three hydrogen atoms and one nitrogen atom.

• The process is continuous. The ammonia is **liquefied** by cooling and then removed. Removing the product drives the reaction forwards.

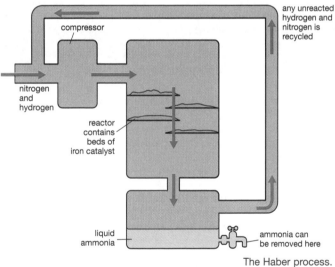

The Haber process.

Questions

1 When hydrogen and nitrogen react to form ammonia, we say that the reaction 'does not go to completion'. What does this mean?

2 What effect does a catalyst have on the amount of product formed in a reversible reaction?

3 What is produced by the Haber process?

4 Name the catalyst that is used in the Haber process.

C2a summary

An **atom** contains **sub-atomic particles**:
- a central **nucleus** of **protons** and **neutrons**
- **electrons**, which orbit the nucleus.

Electrons are arranged around the nucleus in **shells**, which represent **energy levels**: shell 1 has 2 electrons; shell 2 has 8 electrons.

An element's **mass number** = total number of **protons** + total number of **neutrons** per atom. An element's **atomic number** = number of **protons** in an atom.

Electron notation shows the **arrangement** of electrons in the shells in an element, e.g. the notation for magnesium is 2, 8, 2.

Sub-atomic particles

Ionic bonding is the attraction between **oppositely charged ions**.

Non-metal atoms can **share** pairs of electrons to form **covalent bonds**.

Metal atoms **lose** electrons to form **positive** ions. **Non-metal** atoms **gain** electrons to form **negative** ions.

Simple molecular compounds (e.g. water, methane) with covalent bonds:
- do not conduct electricity
- have low melting and boiling points
- are often gases at room temperature
- when solid, are brittle or waxy
- are often insoluble in water.

Ionic compounds:
- are giant ionic **lattices**
- are solids at room temperature (high m.p.)
- conduct electricity when molten or dissolved.

Structure, properties and uses

A **halogen** (Group 7 of the periodic table) exists as a **molecule** of two covalently bonded atoms.

An **alkali metal** (Group 1 of the periodic table) reacts losing one electron to form a positive ion. With non-metals (e.g. halogens), alkali metals form **salts** (ionic compounds).

Giant covalent structures (e.g. diamond, silicon dioxide) are macromolecules that:
- have very high melting points
- are very hard
- do not conduct electricity.

A **nanoparticle** is a tiny structure made with special properties due to the precise way in which its atoms are arranged.
Nanoparticles are used as **biosensors** and may process information, act as catalysts and provide new, very hard, light materials.

Smart materials have particular properties:
- **photochromic** materials react to light
- **thermochromic** materials respond to changes in temperature
- **electroluminescent** materials emit light when an alternating current passes through them.

The **relative formula mass** of a compound or molecule = sum of **relative atomic masses** of all atoms in its **formula**.

High **atom economy** is achieved if a high proportion of the reactants end up as useful products. This is important for **sustainable development** and to limit costs.

A **mole** of a substance is the **relative formula mass** in **grams** of that substance.

Reversible reactions can proceed in both directions. They are affected by: **pressure** and **temperature**, removal of products and presence of a **catalyst**.
Reversible reactions carried out in a 'closed system' will eventually reach **equilibrium**.

Percentage yield can be used to compare the **actual yield** of a **chemical reaction** with the **theoretical yield**.

Composition, yield and equilibrium

The **Haber process** is a reversible reaction:
nitrogen + hydrogen $\rightleftharpoons$ ammonia

The **percentage mass** of an element is given by:

$$\text{\% mass of an element in a compound} = \frac{\text{relative atomic mass of the element} \times \text{no. of atoms of element in the formula}}{\text{relative formula mass}} \times 100$$

Rates of reactions

Measuring the rates of reactions

- Magnesium ribbon reacts with hydrochloric acid, forming magnesium chloride and hydrogen gas.

- The speed of this reaction can be measured by finding:
  - the time taken for the magnesium to dissolve
  - the speed at which hydrogen bubbles form.

- To speed up the reaction, we could:
  - increase the **temperature** or **concentration** of the acid
  - use **powdered** metal
  - or use a **catalyst**.

Magnesium + hydrochloric acid → magnesium chloride + hydrogen.

- The **rate** (speed) **of a chemical reaction** can be measured as the amount of:
  - **reactant** used up in a set period of time:

    $$\text{rate of reaction} = \frac{\text{amount of reactant used up}}{\text{time}}$$

  - **product** formed in a set period of time:

    $$\text{rate of reaction} = \frac{\text{amount of product formed}}{\text{time}}$$

- The rate of reaction between magnesium and an acid can be measured by finding the volume of gas formed (cm^3) in the time taken (seconds). The faster the reaction, the faster the gas fills the syringe.

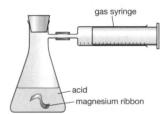

gas syringe

acid

magnesium ribbon

Measuring the volume of gas formed using a gas syringe.

Following the rate of reaction

Methods of measuring reaction rate

- To follow the reaction between magnesium ribbon and hydrochloric acid, we can measure:
  - how quickly it makes hydrogen gas
  - how quickly magnesium loses mass.
  - how long magnesium takes to dissolve

- If a **product** of a reaction is a **gas**,

  (**1**) measure its **volume** by:
  - collecting the gas in an upturned burette or measuring cylinder
  - collecting the gas in a syringe
  - counting bubbles of gas fed into a test tube of water

  (**2**) measure the **mass** of gas lost by finding the mass of the reacting mixture.

- In all these methods, several measurements can be made at timed intervals.

- If a **reactant** is a **solid** that **dissolves**, we can make just one measurement, of the time it takes to disappear.

- For two solutions reacting to form a solid that clouds the solution, measure the time for a cross beneath a flask to disappear. Alternatively measure several times using a sensor and **datalogger**.

Questions

(Grades G-E)

1 State **three** ways in which you could *slow down* the rate of reaction between a piece of magnesium and hydrochloric acid.

(Grades D-C)

2 When following reaction rates, what is a gas syringe used for?

(Grades G-E)

3 A conical flask containing a piece of magnesium and some hydrochloric acid was placed on a top pan balance. The reading on the balance gradually got less. Why did this happen?

(Grades D-C)

4 To find out the rate at which this magnesium ribbon was reacting with the hydrochloric acid in the conical flask, what else would you need to measure?

Collision theory

Increasing the rate of collisions

- A chemical reaction occurs when reacting particles **collide** with **sufficient energy**.

- The minimum energy needed for a reaction to take place is called **activation energy**.

- **Successful collisions** are more likely when there is an *increase* in:
  - the **concentration** of solutions
  - the **pressure** on reacting gases
  - the **temperature** of reactants
  - the **surface area** of reacting solids

  and when there is a **catalyst** for the reaction.

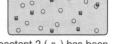

key
- ○ reactant 1
- ▪ reactant 2
- ⊚ successful reaction

- In a solution of two reactants, doubling the concentration of *one* reactant will double the **frequency** of **successful collisions** with the other reactant.

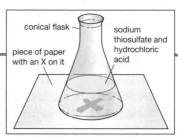

Beaker A   Beaker B

The concentration of reactant 2 (▪) has been doubled in Beaker B.

- The frequency of successful collisions increases when:
  - **pressure** is increased on reactants that are **gases**, because this brings particles closer
  - **temperature** rises, increasing the **kinetic energy** of particles, so they collide more frequently, with greater energy
  - **solids** are broken up (e.g. crushed or powdered), to increase the **surface area** for collisions
  - a **catalyst** is added that weakens bonds in reacting particles.

G–E

D–C

Heating things up

Temperature and reaction rate

- Sodium thiosulfate and hydrochloric acid react to form products that include **solid** sulfur:

  sodium thiosulfate + hydrochloric acid →
  sulfur + sulfur dioxide + sodium chloride + water

  Sulfur clouds the water and it eventually looks opaque, obscuring the cross under the flask, as shown in the diagram.

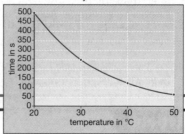

conical flask
piece of paper with an X on it
sodium thiosulfate and hydrochloric acid

Reacting sodium thiosulfate with hydrochloric acid.

- The reaction is repeated with exactly the same reactants but at different temperatures, at 10 °C intervals. The time taken to obscure the cross is recorded for each temperature.

- **Trial experiments** for this reaction show:
  - volumes and concentrations that give a sensible reaction time
  - how to ensure that all **factors** except *temperature* are constant.

- The **graph** for this reaction shows that:
  - temperature increase causes reaction *time* to decrease
  - each 10 °C increase halves reaction time
  - the temperature change 20–30 °C has more effect than 30–40 °C
  - results form a perfect curve, so the results are reliable.

Graph to show the time for a cross to disappear at different temperatures when sodium thiosulfate and hydrochloric acid react.

(graph: time in s vs temperature in °C, y-axis 0–500, x-axis 20–50)

G–E

D–C

Questions

1 Why does increasing the concentration of a reactant increase the rate of a reaction?

2 If you double the concentration of hydrochloric acid in which magnesium is reacting, by how much would this increase the frequency of collisions between particles of reactants?

3 Explain why a cross under a flask becomes invisible when sodium thiosulfate reacts with hydrochloric acid.

4 Use the graph to calculate how much faster the reaction was at 30 °C than at 20 °C.

Grind it up, speed it up

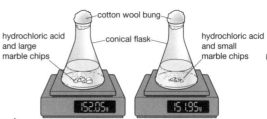

You should be able to: interpret graphs showing the amount of product formed (or reactant used up) with time, in terms of the rate of the reaction.

Surface area and reaction rate

G–E

- Two sizes of marble chips (calcium carbonate) and hydrochloric acid react to form products that include carbon dioxide **gas**:

calcium carbonate + hydrochloric acid → carbon dioxide + calcium chloride + water

As carbon dioxide bubbles off, the mass of the reaction mixture falls. Mass is measured at 30-second intervals.

- You can use **trial experiments** to find out:
  - a mass of marble and a volume and concentration of acid for a reaction lasting about 10 minutes
  - how much acid to use so that all the magnesium reacts
  - how to ensure all **factors** except the *size* of the marble chips are constant.

D–C

- The **graph** shows that: the total volume of gas is the same in both experiments; the reaction rate is highest at the start.

- We can use the 1-minute reading to find the **initial reaction rate** as *mass* of gas per minute:

rate of reaction = $\dfrac{\text{amount of product formed}}{\text{time}}$

- The rate of reaction is greater for the smaller chips than the larger chips.

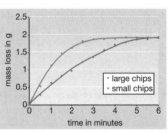

Finding the rate of loss in mass when marble chips of two different sizes react with acid.

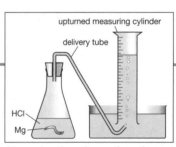

Graph to show the mass of carbon dioxide gas lost when acid reacts with marble chips of two different sizes.

Concentrate now

Concentration and reaction rate

G–E

- Magnesium reacts with sulfuric acid to form products that include hydrogen gas which can be collected:

magnesium + sulfuric acid → hydrogen + magnesium sulfate

An upturned measuring cylinder collects the hydrogen.

- Gas volume is recorded at 1-minute intervals. The reaction is repeated for five concentrations of acid, with all other conditions constant.

D–C

- **Trial experiments** can be used to show:
  - masses of magnesium and concentrations of sulfuric acid that give sensible reaction times
  - how to ensure all **factors** except acid *concentration* are constant.

- The **graph** shows that: increase in acid concentration causes rate of reaction to increase; the reaction rate is highest at the start.

Measuring the volume of gas formed when magnesium reacts with different concentrations of acid.

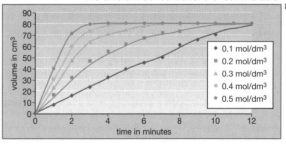

Graph to show the volume of hydrogen formed with five different concentrations of sulfuric acid.

Questions

(Grades G-E)
1 Marble reacts with hydrochloric acid to produce hydrogen. Which would produce hydrogen faster – lumps of marble or tiny marble chips (assuming the total mass of marble is the same)?

(Grades D-C)
2 Explain why measuring the mass of gas lost after 5 min would not give you useful information about the rate of the reaction between hydrochloric acid and marble chips.

(Grades G-E)
3 Name the gas produced when magnesium reacts with hydrochloric acid.

(Grades D-C)
4 Explain why using 2 mol/dm³ acid rather than 1 mol/dm³ of acid would double the rate of reaction.

Catalysts

Effect and behaviour of catalysts

- Catalysts speed up reaction rates, or allow reactions to take place that would otherwise not happen. Catalysts are not used up in reactions. Different reactions have different catalysts.

- Hydrogen peroxide (H_2O_2) decomposes slowly to water and oxygen, but very quickly with manganese(IV) oxide used as a catalyst.

  hydrogen peroxide → water + oxygen

- Burning petrol produces the poisonous gas carbon monoxide, and oxides of nitrogen that contribute to **acid rain**. **Catalytic converters** make them react to produce less harmful gases:

  carbon monoxide + nitrogen dioxide → carbon dioxide + nitrogen

- Weighing manganese(IV) oxide before and after the reaction proves that it is not used up when hydrogen peroxide decomposes.

- Catalysts include **transition metals** (e.g. manganese) and their compounds.

- In the **Haber process**, an **iron** catalyst helps nitrogen and hydrogen gases to react to form ammonia gas.

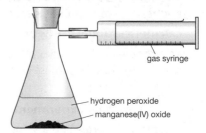
gas syringe

hydrogen peroxide
manganese(IV) oxide

The decomposition of hydrogen peroxide can be followed by measuring the volume of oxygen that forms.

How Science Works

You should be able to: explain and evaluate the development, advantages and disadvantages of using catalysts in industrial processes.

G–E

D–C

Energy changes

Exothermic and endothermic reactions

- In a reaction, **energy** is either given out (transferred *to* the surroundings), or taken in (transferred *from* the surroundings).

- During **exothermic** reactions, energy is given out, often as heat. Examples include all **combustion** reactions, many **oxidations**, and **neutralisations** (e.g. hydrochloric acid + sodium hydroxide → sodium chloride + water).

- During **endothermic** reactions, energy is taken in, often as heat. Examples are **thermal decompositions** (e.g. calcium carbonate → calcium oxide + carbon dioxide).

G–E

Energy changes in reversible reactions

- A **reversible reaction** that is **exothermic** in one direction is **endothermic** in the opposite direction. In both directions the same amount of energy is transferred.

- When water is added to **anhydrous copper sulfate** (see page 98), the white powder turns blue and *gives out* a lot of heat. This reaction (a test for water) is exothermic.

  exothermic
  anhydrous copper sulfate + water ⇌ hydrated copper sulfate
  endothermic

- When hydrated copper sulfate is heated, the crystals *take in* an equal amount of heat that drives off the water. This reaction is endothermic.

D–C

Questions

Grades G-E

1 Name a catalyst that can speed up the breakdown of hydrogen peroxide to water and oxygen.

Grades D-C

2 Name the group of metals that make good catalysts.

Grades G-E

3 When hydrochloric acid reacts with sodium hydroxide in a test tube, the tube gets hot. Is this an exothermic reaction, or an endothermic reaction?

Grades D-C

4 What can you say about the amount of energy taken in or given out in a reversible reaction?

Equilibrium 2

Reaching a balance point

G–E

- When heat breaks down calcium carbonate in a closed system (no products can enter or leave), there is a **reversible reaction**:

  calcium carbonate ⇌ calcium oxide + carbon dioxide (See page 98.)

- In time, the reverse reaction balances the forward reaction. This balance point is called the **equilibrium**.

D–C

- Increasing the heat breaks down more calcium carbonate, and less of the calcium oxide and carbon dioxide re-form calcium carbonate. **Changing a condition** (temperature) has moved the equilibrium forwards. (See page 99.)

Industrial processes

The Haber process

G–E

- **Ammonia** is required to manufacture **nitrate** fertilisers. Ammonia is made from nitrogen and hydrogen in a reversible reaction (see also page 99):

  $$nitrogen + hydrogen \rightleftharpoons ammonia$$
  $$N_2(g) + 3H_2(g) \rightleftharpoons 2NH_3(g)$$

  number of molecules: 1 nitrogen   3 hydrogen   2 ammonia

- Conditions are:
  - temperature 450 °C
  - pressure 200 atmospheres
  - iron catalyst.

D–C

- **Yield** is 30%. This means that 30% of the starting materials are converted to product.

- This is acceptable because the **reaction rate** (see page 101) is good.

- Also, costs are relatively low. Reducing energy costs contributes to **sustainable development**.

Conditions for the Haber reaction

D–C

- For a good **yield**:
  - the reaction is **exothermic**, so low temperature increases yield; however, this lengthens the time to reach equilibrium
  - high pressure drives the reaction forwards.

- For a good **rate** but lower yield, higher temperatures are used to produce ammonia faster.

How Science Works

You should be able to: describe the effects of changing the conditions of temperature and pressure on a given reaction or process.

Questions

Grades G-E

1 A reversible reaction eventually settles down so that the reverse reaction balances the forward reaction. What name is given to this point?

Grades D-C

2 Look at the equation for the thermal decomposition of calcium carbonate. How could we move the equilibrium *backwards*?

Grades G-E

3 Ammonia is made in the Haber process. Give **one** use of this ammonia.

Grades D-C

4 State the conditions under which the reactions in the Haber process take place.

Free ions

Freeing ions in a solid

- Ionic compounds are solids. Their **ions** are arranged in a **giant lattice** (see page 92).

- Ions are free to move when the solid is **dissolved** in water to form a solution, or is **melted**.

- Heat melts solid lead bromide:

$$PbBr_2(s) \rightarrow PbBr_2(l)$$

- Passing an electrical current through the molten lead bromide splits it into its **elements**, molten lead and bromine gas:

$$PbBr_2(l) \rightarrow Pb(l) + Br_2(g)$$

This process is **electrolysis** (see also page 31).

- (s) = solid, (l) = liquid, (g) = gas. Also, (aq) = aqueous (dissolved in water).

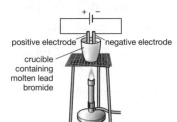

positive electrode — negative electrode

crucible containing molten lead bromide

To free the ions, an ionic compound is kept molten. Then, it can be electrolysed to separate its elements.

Electrolysis of ionic compounds

- Ionic compounds always contain **positive metal** ions and **negative non-metal** ions. They are solids at normal temperatures, so the ions are not free to move.

- They can be electrolysed when either molten or in solution.

- In electrolysis, the ions move towards the *oppositely* charged electrode. Positive metal ions move to the **negative** electrode, and negative non-metal ions move to the **positive** electrode.

positive electrode — negative electrode

The ions in molten sodium chloride move to the electrode with a charge opposite to theirs.

Electrolysis equations

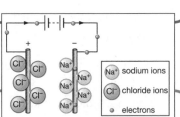

Na⁺ sodium ions
Cl⁻ chloride ions
electrons

The sodium ions (Na^+) and chloride ions (Cl^-) move to the oppositely charged electrodes.

Electrolysis of molten sodium chloride

- Sodium ions are positively charged, so they are attracted to the negative electrode. Chloride ions are negatively charged, so they are attracted to the positive electrode.

- The diagram shows what happens to the sodium ions and chloride ions at the electrodes.

Each sodium ion gains an electron and each chloride ion loses an electron.

- At the negative electrode, a positively charged sodium **ion** *gains* an electron: it is **reduced**.

  – It becomes an uncharged atom of sodium **metal**.

- At the positive electrode, a negatively charged chloride ion *loses* an electron: it is **oxidised**.

  – It becomes an uncharged chlorine atom. It shares electrons with another chlorine atom to form a molecule of chlorine **gas**.

- Page 107 describes the uses made of the products from the electrolysis of sodium chloride.

Questions

(Grades G-E)
1 In which **two** states can lead bromide be electrolysed?

(Grades D-C)
2 Explain why lead ions go to the negative electrode when lead bromide is electrolysed.

(Grades G-E)
3 A student wrote this sentence: 'Sodium is positive so it goes to the negative electrode.' Decide what is wrong, and write the sentence correctly.

(Grades D-C)
4 Describe what happens to chloride ions at the positive electrode. Are they reduced or oxidised?

Uses for electrolysis

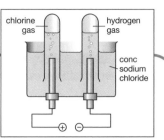

Electrolysing sodium chloride (salt) solution.

Sodium chloride and copper solutions

- A **solution** of sodium chloride contains Na^+ and Cl^- ions from salt, and H^+ and OH^- ions from water.

- The **reactivity series** (see below) arranges metals and hydrogen in order of their reactivity.

- In **electrolysis**, H^+ and Na^+ ions go to the negative electrode. Hydrogen forms, but not sodium; it is more reactive than hydrogen.

- The electrolysis **products** are the gases chlorine and hydrogen. Na^+ and OH^- remain in solution as ions of **sodium hydroxide**.

- **Chlorine** is used to make bleach, chlorinated solvents and PVC. **Hydrogen** is a fuel, and is used to make ammonia and to change vegetable oils to margarine. Sodium hydroxide is used to make soap, detergents, paper and to purify bauxite (aluminium ore).

- To **purify copper** metal, a piece of the impure copper is used as the positive electrode in a solution containing copper ions, Cu^{2+}.

- Copper ions leave the positive electrode and move through the solution to the **pure** copper *negative* electrode. They gain two electrons and become Cu atoms, building up on the electrode.

- More reactive impurities, including zinc, stay in solution. Unreactive impurities collect as sludge.

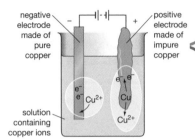

Copper transfers from the positive to the negative electrode.

> **How Science Works**
>
> You should be able to: predict the products of electrolysing solutions of ions.

Acids and metals

> **Top Tip!**
>
> If you are asked whether a metal reacts with acids, look at the reactivity series.

Which metals react with acids to make salts?

- Any **metal** *higher* than hydrogen in the **reactivity series** can react with an acid. The reactive metal **displaces** the less reactive hydrogen in the acid. A soluble **salt** is formed, and hydrogen is given off.

- Potassium and sodium are high in the series. They are too reactive to react safely with sulfuric acid. Copper and silver are lower than hydrogen and they will not react with dilute sulfuric acid.

- Salts from acids: sulfuric acid → sulfates; nitric acid → nitrates; hydrochloric acid → chlorides.

- A **salt** is formed when hydrogen in an acid is replaced by a metal or an **ammonium** ion.

- To make zinc sulfate, the **reagents** are zinc metal and sulfuric acid:

  zinc + sulfuric acid → zinc sulfate + hydrogen

  Zinc is added in **excess** to ensure that all the acid is used up and the solution contains only zinc sulfate. The solution is filtered to remove any remaining zinc. The water in the solution is evaporated to leave zinc sulfate crystals.

Reactivity series	
K	**more reactive**
Na	
Mg	
Al	
Zn	
Fe	
Pb	
H	
Cu	
Ag	**less reactive**

> **Top Tip!**
>
> Learn the formulae of these acids:
> sulfuric acid = H_2SO_4
> nitric acid = HNO_3
> hydrochloric acid = HCl

Questions

1 Name the gas that is formed at the negative electrode when sodium chloride solution is electrolysed.

2 Give **two** uses of this gas.

3 You have a piece of zinc. What acid would you add it to, to make zinc chloride?

4 Explain why you need to filter the mixture when the reaction finishes, if you want to get pure zinc chloride.

Making salts from bases

Bases, alkalis and reaction with acids

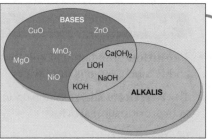

All the substances are bases, but only some are alkalis.

- **Bases** react with acids to form **salts**. Oxides and hydroxides of metals are bases. Soluble bases are called **alkalis**.

- **base + acid → salt + water**

- Solid zinc oxide reacts with warm acid:

  zinc oxide + hydrochloric acid → zinc chloride + water

- Unreactive metals will not react with acid, but their oxide (or hydroxide) *will* react.

Making a salt from an acid and an insoluble base

- Magnesium oxide and hydrochloric acid react to form magnesium chloride salt:

  magnesium oxide + hydrochloric acid → magnesium chloride + water

  $MgO(s)$ $+ 2HCl(aq)$ $→ MgCl_2(aq)$ $+ H_2O(l)$

- Adding **excess** oxide to warmed acid ensures all the acid reacts. So the solution contains only magnesium chloride. The excess magnesium oxide can be removed by filtering. Evaporating water in the solution leaves magnesium chloride powder.

Acids and alkalis

Using indicators to follow reactions of acids and alkalis

- The colour of an **indicator** shows how acid or alkaline a solution is – its **pH**.

- You can add indicator to an alkali and then gradually add acid to it. The indicator will change colour when just enough acid has been added to react with all the alkali. This is the point of **neutralisation**.

- At neutralisation, the solution contains a **salt** and water only.

- **Ammonia** dissolves in water to give an alkaline solution. This can be reacted with acids to produce ammonium salts, which are important as fertilisers.

Making a salt from an acid and an alkali

- **Phenolphthalein** is a good indicator to follow the reaction of sodium hydroxide with hydrochloric acid to form sodium chloride (common salt).

  sodium hydroxide + hydrochloric acid → sodium chloride + water

  pH   14   1   7

  At pH 14, phenolphthalein is pink.
  At pH 7 or below, it is colourless.

- The diagrams show the procedure.

- If pure salt is required, the procedure is repeated using the same concentrations and volumes but without the indicator. The water is then evaporated to produce solid salt.

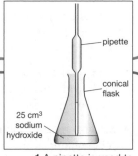

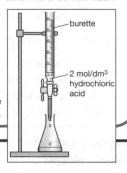

1 A pipette is used to transfer exactly 25 cm³ of 2 mol/dm³ hydrochloric acid into a conical flask.

2 Two or three drops of phenolphthalein indicator are added to the flask. Because the solution is alkaline, it goes pink.

25 cm³ sodium hydroxide

phenolphthalein

3 Hydrochloric acid is added until one drop of the hydrochloric acid turns the indicator colourless.

pipette
conical flask
25 cm³ sodium hydroxide
burette
2 mol/dm³ hydrochloric acid

Questions

Grades G-E
1 What products would you get if you add magnesium oxide to hydrochloric acid?

Grades D-C
2 Name an oxide and an acid that would react to produce zinc sulfate.

Grades G-E
3 What does an indicator tell you about a liquid?

Grades D-C
4 Explain why, if you want to make a pure sample of sodium chloride, you would first react the base and acid using an indicator, and then do it again *without* the indicator.

Neutralisation

What are acids and alkalis?

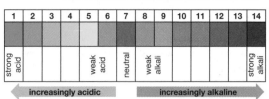

Top Tip!

Acids are substances that contain $H^+(aq)$.
Alkalis are substances that contain $OH^-(aq)$.

- All **acids** contain hydrogen. In water, it is released as hydrogen ions, $H^+(aq)$.

- All **alkalis** contain hydroxide, released in water as hydroxide ions, $OH^-(aq)$.

- The **strength** of an acid or alkali depends on the amount of H^+ or OH^- ions it releases in water.

1	2	3	4	5	6	7	8	9	10	11	12	13	14
strong acid				weak acid		neutral	weak alkali						strong alkali

← increasingly acidic   increasingly alkaline →

Universal indicator: the pH scale shows how strong an acid or alkali is.

- **Universal indicator** goes red in a strong acid, indicating a pH of 1. In a weak acid (e.g. vinegar), pH is 3.

- In **neutralisation** reactions, the hydrogen and hydroxide ions react to form water:

$$H^+(aq) + OH^-(aq) \rightarrow H_2O(l)$$

1	2	3	4	5	6	7	8	9	10	11	12	13	14
digestive juices in your stomach		lemon juice		orange juice		rain water	pure water	saliva	sea water	sodium hydrogencarbonate	ammonia	limewater	sodium hydroxide

Most solutions around us are not neutral.

← increasingly acidic increasing concentration of $H^+(aq)$ increasingly alkaline increasing concentration of $OH^-(aq)$ →

Precipitation

Forming a precipitate, and hard water

- **Insoluble salts** can be made by mixing two soluble salts that contain the appropriate ions. If the ions react to produce an insoluble salt, it forms a solid **precipitate** that can be separated.

- To form the insoluble yellow compound lead chromate, compounds containing lead ions and chromate ions are required, for example:

  lead nitrate + potassium chromate → potassium nitrate + lead chromate

- **Hard water** contains dissolved **calcium** ions. To remove them, adding soluble sodium **carbonate** forms insoluble calcium carbonate that precipitates out:

  soluble calcium ion + soluble carbonate ion → insoluble calcium carbonate

Water and effluent treatment

- Before treatment, the water supply contains **nitrate** ions from fertilisers and **phosphate** ions from washing powders.

- Phosphate ions are removed using calcium, iron or aluminium ions:

  soluble calcium ion + soluble phosphate ion → insoluble calcium phosphate

- **Heavy metals** are removed as insoluble **carbonates**.

- Several **metals** can be identified from their insoluble **hydroxides** which have characteristic colours. To form the precipitate, sodium hydroxide is added to water containing the metal ions.

How Science Works

You should be able to:
- suggest methods to make a named salt
- explain and evaluate processes that use the principles described in this unit.

Questions

(Grades G-E)
1 Write the formula of the ion that is present in all acids.

(Grades D-C)
2 Write an equation to show what happens when any acid reacts with any alkali.

(Grades G-E)
3 What is a 'precipitate'?

(Grades D-C)
4 Explain why adding calcium ions to water can remove phosphate ions from it.

C2b summary

Collision theory states that **particles** must collide with sufficient energy in order to react.

The **minimum energy** required for a reaction to take place (a **successful collision**) is the **activation energy**.

Rates of reaction

The **rate of a reaction** can be measured as:
- the amount of **reactant used up** in a set time
- or the amount of **product formed** in a set time.

Reactions can be **speeded up** by **increasing** the:
- **temperature**
- **concentration** of a solution
- **pressure** of a gas
- **surface area** of a solid

or by using a **catalyst**.

Exothermic reactions give OUT energy.
Endothermic reactions take IN energy.

If a **reversible reaction** is exothermic in one direction, it is endothermic in the opposite direction, e.g.

anhydrous copper sulfate + water

exothermic

⇌ **hydrated copper sulfate**

endothermic

Types of reaction

In reversible reactions, **equilibrium** is reached at a point when the rate of the reverse reaction **balances** the rate of the forward reaction.

The conditions used for the **Haber process** are:
- temperature 450 °C
- pressure 200 atmospheres
- iron catalyst.

When **molten** or **dissolved** in water, the **ions** in **ionic compounds** are **free** to move.

If ions are free to move, passing an **electric current** through an ionic compound breaks it down into its **elements**. This is called **electrolysis**.

Electrolysis of **sodium chloride** (common salt) solution makes hydrogen, chlorine and sodium hydroxide. **Copper** can be purified by electrolysis.

Ions

During **electrolysis**:
- **negative** non-metal ions **lose electrons** at the positive electrode to form atoms and molecules
- **positive** metal ions **gain electrons** at the negative electrode to form atoms.

OIL RIG:
Oxidation **I**s **L**oss
Reduction **I**s **G**ain – of electrons.

Metal oxides and hydroxides are **bases** and react with **acids** to form **salts**.

Soluble bases are **alkalis** and these react with acids to make salts.

In **neutralisation** reactions, hydrogen ions and hydroxide ions react to form water:
$$H^+(aq) + OH^-(aq) \rightarrow H_2O(l)$$

Some **metals** can be reacted with **acids** to make **salts**. The **reactivity series** of metals can be used to find out if a metal will react with an acid.

Making salts

Salts from acids:
- sulfuric acid → sulfates
- nitric acid → nitrates
- hydrochloric acid → chlorides.

Insoluble salts can be made as **precipitates** when two solutions are mixed together.

To remove **ion impurities** from the **water supply**, they can be made into **insoluble salts**, e.g. phosphate ions from washing powders can be removed using calcium ions to form insoluble calcium phosphate.

See how it moves!

Distance-time graphs

- A **distance-time graph** shows the **distance** an object moves in a period of **time**. The **slope** (distance/time) shows its **speed**.

- The graph shows Emma's journey to school:
  - A–B: walks between home and bus stop; *shallow* slope = *slow* speed
  - B–C: waits for bus; *horizontal* line = staying in the same place, **stationary**
  - C–D: on the bus; *steep* slope = *fast* speed
  - D–E: still on the bus; less steep slope = bus speed slower
  - E–F: off at the bus stop near the school, waits for a friend; horizontal line = stationary.

- A distance-time graph does not show whether a journey is straight or changes direction.

Emma's journey to school.

Speed and average speed

- Speed on a journey may vary. For the total journey:

  average speed = $\dfrac{\text{total distance travelled}}{\text{total time taken}}$

Top Tip!
A distance-time graph that slopes downwards shows an object coming back towards its starting point, not one that is slowing down.

Speed isn't everything

Velocity involves speed and direction

Top Tip!
If a question does not tell you which way velocity is being measured, you can choose which way is positive, but keep it the same for the whole question.

- **The velocity, *v*, of an object is its speed in a given direction.**

- The diagrams show the velocities of three objects in an *upward* direction.

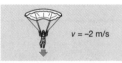

The **positive velocity** of the rocket tells us that it is moving at 12 000 m/s upwards, away from the ground.

The parachute has a **negative velocity** because it is travelling **towards** the ground and velocity is measured **away** from the ground.

The aeroplane might have a very high speed, but its velocity away from the ground is 0 m/s. It is not moving towards or **away from the ground**.

How Science Works
You should be able to: construct distance-time graphs for a body moving in a straight line when the body is stationary or moving with a constant speed.

Acceleration and direction

- Any **change** in velocity, whether speeding up or slowing down, is called **acceleration**. The value is negative for slowing down.

  acceleration (in m/s^2) = $\dfrac{\text{change in velocity (in m/s)}}{\text{time taken for change (in s)}}$

- Like velocity, acceleration must have a direction.

- The archer's arrow accelerates towards the right.

The arrow is fastest when it leaves the bow, so its acceleration is negative.

Questions

(Grades G-E)
1 Look at the distance-time graph for Emma's journey to school. At which stage was she travelling fastest? How can you tell?

(Grades D-C)
2 Use the average speed equation to calculate Emma's average speed for her journey to school.

(Grades G-E)
3 Explain the difference between 'speed' and 'velocity'.

(Grades D-C)
4 A stationary car started to move, and reached a speed of 20 m/s in 20 seconds. Use the acceleration equation to calculate its acceleration in m/s^2.

Velocity-time graphs

Velocity-time graphs

- A **velocity-time** graph shows how the **velocity** of an object changes with time.

- The graph shows the changing velocity of a cyclist. A horizontal line indicates **constant velocity**.

- The **slope** (gradient) of a velocity-time graph represents **acceleration**.

- Steep slope A on the cyclist's graph (1) is a positive slope, showing a positive acceleration.

- Shallower slope C is a negative slope, indicating a negative acceleration.

- The area under a velocity-time graph (2) shows the **distance** an object travels.

- The graphs (3) are for a skydiver who:
  - A: leaves plane; velocity increases
  - B: reaches maximum velocity
  - C: opens parachute; slows rapidly
  - D: reaches constant, lower velocity
  - E: hits ground.

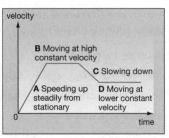

(1) Velocity-time graph for a cyclist.

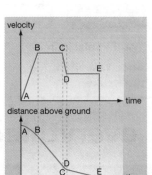

Top Tip!

When you describe the line on a velocity-time graph showing constant velocity, say that the line is 'horizontal', not just 'straight'.

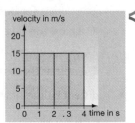

(2) The area under the graph shows the distance a model plane travels.

Top Tip!

The slope (gradient) of a velocity-time graph shows the acceleration.

(3) Velocity-time graph and distance-time graph for a skydiver.

Let's force it!

Adding forces

- All the different forces on an object act together as a single, **resultant force**. **The resultant force has the same effect as all the forces acting on the object.**

- Forces acting in the *same* direction are added. Forces acting in the *opposite* direction are subtracted. For the car:

  forward force – air resistance – friction = 1000 N – 400 N – 400 N

The resultant force acting on the car moves it forwards.

- If the resultant force is zero:
  - **stationary** objects remain stationary
  - moving objects maintain the same **speed**, in the same **direction**.

- If the resultant force is not zero:
  - stationary and moving objects start to **accelerate** in the direction of the resultant force.

Questions

(Grades G-E)

1 Look at the velocity-time graph for the cyclist. Describe what the line on the graph looks like when he is cycling at a constant speed.

(Grades D-C)

2 Sketch a velocity-time graph to show a car moving at a steady speed, slowing down and stopping at traffic lights, then accelerating back up to the same steady speed.

(Grades G-E)

3 A parachute is falling through the air. The downwards force of gravity on it is exactly the same as the upwards force of air resistance. What can you say about its velocity? Explain your answer.

(Grades G-E)

4 Tom is trying to push a box along the ground. He pushes with a force of 10 N. Friction between the box and the ground exerts a force of 10 N in the other direction. What happens to the box?

Force and acceleration

Unbalanced forces

- When a car travels at a **steady** (constant) **speed**, the forwards **driving force** exactly balances the backwards **frictional forces** of air and road.

- **Acceleration** occurs in the *direction* of the resultant force:
  - accelerating increases the driving force, increasing speed
  - braking increases the frictional force, reducing speed.

- For the same mass, a bigger force gives a bigger acceleration.

- For the same force, a bigger mass gives a smaller acceleration.

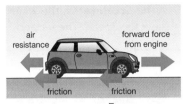

Forces on a car.

Measuring acceleration

- Different **forces** and different **masses** are used with the apparatus shown. In each trial, the **force** is made constant:
  - the **force meter** measures **force** (newtons)
  - the **ticker-timer** measures **velocity** (metres/second).

ticker-timer machine   lab trolley   force meter

Apparatus to investigate how acceleration changes when force changes.

- Velocity-time graphs drawn from the results show that **acceleration** is:
  - given by the **slope** (velocity/time)
  - **directly proportional** to force, $a \propto F$
  - inversely proportional to mass, $a \propto 1/m$

- **resultant force (N) = mass (kg) × acceleration (m/s²): $F = ma$**

Top Tip!

The bigger the force, the bigger the acceleration. The bigger the mass, the smaller the acceleration.

Balanced forces

Balanced forces and stationary objects

- **For every force there is an equal and opposite force**.

- When two forces are balanced, they have the *same* value but act in *opposite* directions. Examples for **stationary** objects are:
  - the tug-of-war teams (see picture)
  - a person floating in water; the **upthrust** of the water balances the downwards force of the person's weight.

The two teams are stationary, pulling with equal but opposite force. The forces are balanced.

Forces and moving objects

- Objects moving at **constant velocity** also have balanced forces.

- When a gun is fired, the force of explosive drives:
  - the bullet forwards from the gun at high speed
  - the gun backwards a little, at lower speed.

air resistance

weight

When a parachutist is falling at constant velocity, his weight acting downwards balances the **air resistance** acting upwards.

Questions

Grades G-E

1 A small mass and a large mass are each pushed with the same force. Which one accelerates more?

Grades D-C

2 Complete this sentence: Acceleration is .... proportional to force.

Grades G-E

3 Complete this sentence: If equal and .... forces act on a stationary object, it does not move.

Grades D-C

4 Complete this sentence: If equal and opposite forces act on a moving object, it does not ....

Terminal velocity

Reaching terminal velocity

- To walk into a strong wind, a man exerts a **resultant force** greater than the forces of **wind resistance** and **friction** between shoes and ground. As he **accelerates**, the wind resistance increases. He reaches a steady slow top speed. Then, opposing forces balance.

- A falling object always reaches a top speed, its **terminal velocity**. At this speed, the force of **weight** downwards cancels the force of **air resistance** upwards.

G–E

Terminal velocity graph

- As a skydiver falls through the air, she accelerates. As she gets faster, air resistance increases. Eventually, air resistance exactly balances the downwards force of her weight. She reaches her terminal velocity.

- The parachute opens. Air resistance increases, exerting a force greater than the skydiver's weight. The resultant force is upwards and she slows down until weight = air resistance at a lower terminal velocity than before.

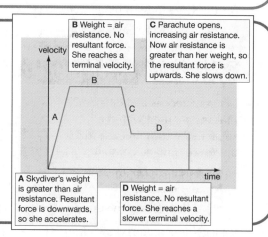

B Weight = air resistance. No resultant force. She reaches a terminal velocity.

C Parachute opens, increasing air resistance. Now air resistance is greater than her weight, so the resultant force is upwards. She slows down.

A Skydiver's weight is greater than air resistance. Resultant force is downwards, so she accelerates.

D Weight = air resistance. No resultant force. She reaches a slower terminal velocity.

Velocity-time graph for a skydiver.

D–C

Stop!

Stopping distance

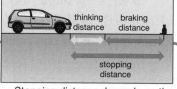

Stopping distance depends on the speed of the car.

- A driver sees a cat in the road. Her car moves a **thinking distance** before she brakes. Then the car moves a **braking distance** before halting: **stopping distance = thinking distance + braking distance**

- A moving car has kinetic energy. To stop the car, this kinetic energy must be reduced to zero. This is done by the braking force.

- The greater the car's speed, the greater the braking force needed to stop it within a particular distance.

G–E

Thinking distance and braking distance

- A driver's **reaction time** and the car's **speed** both affect the thinking distance:
thinking distance = speed × reaction time

- Drinking, taking drugs or being tired prolongs the reaction time.

- **Braking distance** depends on:
  - **speed squared**; doubling the speed increases the braking distance four times
  - the car's **mass**; kinetic energy is proportional to mass
  - **friction** of brakes and tyres; greater friction shortens the braking distance, icy or wet roads lengthen it.

How Science Works

You should be able to:
- draw and interpret velocity-time graphs for bodies that reach terminal velocity, including a consideration of the forces acting on the body
- calculate the weight of a body using:
weight (N) = mass (kg) × gravitational field strength (N/kg)

D–C

Questions

Grades G-E

1 What is meant by the 'terminal velocity' of a falling object?

Grades D-C

2 Explain how a parachute reduces the terminal velocity of a person falling through the air.

Grades G-E

3 Complete this equation: stopping distance = ... distance + ... distance.

Grades D-C

4 State **two** factors that might *increase* the thinking distance.

Moving through fluids

Forces in fluids

- Liquids and gases are **fluids** because they flow.

- An object moving through a fluid pushes fluid particles aside and rubs against them.

- **Frictional** forces between fluid and object include **air resistance** and **drag** in water. They increase as:
  - the **speed** of a moving object increases
  - the fluid gets **denser**, so an object displaces a greater **mass** of fluid to move the same distance.

- An object falling through a liquid accelerates due to the force of gravity, until it reaches a **terminal velocity**. Then, the resultant force is zero and:

  upthrust (liquid's resistance force) = object's weight due to gravity

The cyclist must apply greater force than on the road because the drag force of water is greater than air resistance.

Maximising speed through fluids

- Water resistance (drag) slows objects much more than air resistance does.

- At high speed, a jet ski lifts up higher in the water, reducing water resistance.

- Birds and fish have a **streamlined** shape to reduce air or water resistance and so minimise the energy they require to move.

Energy to move

Kinetic energy

- **Kinetic energy** is the energy of **movement**.

- Kinetic energy comes, for example, from: chemical energy in food and in petrol; electrical energy to a motor.

- Kinetic energy is **transformed** when: a hammer hits a horseshoe and **does work** flattening it; a gong is hit and produces **sound**; rubbing your hands produces **heat**.

- **Energy transfer diagrams** show how energy is transferred and transformed.

| kinetic energy in dynamo | → | electrical energy in wires |

bicycle dynamo

| chemical energy in battery | → | electrical energy in wires | → | kinetic energy in toy car |

toy car

There is one energy transfer for a bicycle dynamo. There are two energy transfers for the toy car. In each stage, energy is also transformed.

Friction, heating and transformation

- A car's **kinetic energy** is transformed into **work** to overcome **frictional forces**.

- Much of the **work** against frictional forces is transformed into **heat**.

- A swing continually gains height and loses speed, then gains speed and loses height. **Kinetic energy** is transformed into **potential energy** and back again. The swing slows down because of friction between the swing and air.

How Science Works

You should be able to: discuss the transformation of kinetic energy to other forms of energy in particular situations.

Questions

(Grades G-E)
1 What is the difference between 'air resistance' and 'drag'?

(Grades D-C)
2 Why do fish need a more streamlined shape than birds?

(Grades G-E)
3 A builder drops a brick from the top of a wall. What happens to the kinetic energy in the brick when it hits the ground?

(Grades D-C)
4 A child plays on a swing. At which point on the swing's movement does it have: **a** most potential energy; **b** most kinetic energy?

Working hard

Work and energy

- When a **force** moves an object, **work** is done and **energy** is transferred:

  work done = energy transferred

- In the drawings:
  - the man does work against **air resistance**
  - the gardener does work against **friction**
  - the dog owner does work against **gravity**, transforming muscle energy into **potential energy** in the dog he lifts.

Each person is doing work by moving or lifting something.

How much work?

- Pushing a heavy trolley 10 metres requires greater **force**, transferring more energy, than pushing a light trolley.

- Pushing a trolley 20 metres requires more work than pushing it 10 metres.

- Amount of **work done** depends on the force and on the **distance**:

  work done = force × distance moved in the direction of the force
  (joules, J)   (newtons, N)   (metres, m)

Both men move their blocks the same amount. But the man below also does work against the friction between the ramp and the block.

How much energy?

Kinetic energy

- An object's **kinetic energy** increases when its:
  - **mass** increases
  - **speed** increases.

- All the **kinetic energy** of an object is transferred if it collides with another object and then stops. If it continues to move, but at a slower speed, then only some of its kinetic energy has been transferred.

Potential energy and elastic potential energy

- The **potential energy** of the top trolley is **transformed** into kinetic energy as it runs down the ramp.

- **Work done** to stretch the elastic band behind the trolley is stored as **elastic potential energy**. This is transformed into kinetic energy when the trolley moves forwards.

- Work done to blow up the balloon, stored as elastic potential energy, is transformed into kinetic energy when the trolley moves.

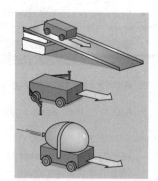

Three energy transformations.

Questions

(Grades G-E)
1 What are the units in which work is measured? Choose from: joules, newtons, grams.

(Grades D-C)
2 Jake lifts two parcels, each weighing 15 N, onto a desk 1 m above the ground. How much work does he do?

(Grades G-E)
3 A lorry and a motorbike are both moving at the same speed. Which one has more kinetic energy, and why?

(Grades D-C)
4 Kay uses an elastic band as a catapult to fire a paper pellet across the room. Where did the kinetic energy in the pellet come from?

Momentum

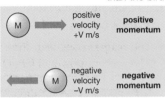

The faster and more massive train has greater momentum than the bird.

What is momentum?

Top Tip!

Kinetic energy is a scalar quantity; momentum is a vector quantity.

- **Moving** objects have **momentum**.
  Momentum depends on an object's **mass** and its **velocity**:

  momentum = mass × velocity
     (kg m/s)   (kg)   (m/s)

- Momentum increases as mass and velocity increase.

- Momentum indicates how hard it is to halt a moving object.

- Momentum is a **vector** quantity – it has **size** (**magnitude**) and **direction**. Depending on direction, its value is positive or negative (see diagram opposite).

- Momentum is *not* the same as **kinetic energy**. Kinetic energy is a **scalar** quantity – it has size but not direction. Its value is always positive.

G–E

The two identical masses move in opposite directions. One has a positive momentum, the other a negative momentum.

Conservation of momentum

How Science Works

You should be able to: use the conservation of momentum to calculate the mass, velocity or momentum of a body involved in a collision or explosion.

- To investigate **momentum**, two trolleys of known **masses** collide. Ticker tape gives their **velocities** immediately before and after collision.

An inelastic collision.

D–C

- The **Law of Conservation of Momentum** states that:

  total momentum before collision = total momentum after collision

  – This applies as long as *no external forces* act on the bodies.

- In an inelastic collision, a 1 kg mass collides at 10 m/s with a stationary 4 kg mass:

before	after	rearranged
$(1 \times 10) + (4 \times 0)$	$= (1 \times v) + (4 \times v)$	$10 + 0 = 5v$
		$v = 10/5 = 2\,\text{m/s}$

Off with a bang!

total momentum = 0   total momentum = $+mv -mv = 0$

momentum = $-mv$ momentum = $+mv$

before   after

Explosions and jet engines

- Before an object **explodes**, nothing moves – momentum is zero. When it explodes, pieces fly in all directions. In any direction, the net momentum of the pieces will be zero. Momentum is conserved.

G–E

Identical masses explode apart. Positive momentum = negative momentum, so total momentum is zero.

- In an aeroplane, the jet engine blasts gas backwards and the aeroplane moves forwards.

Examples of momentum

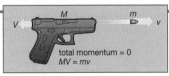

total momentum = 0
$MV = mv$

A lighter gun would recoil more.

- All beads in a Newton's cradle have the same **mass**. One bead hitting one end transfers just enough momentum to make one bead move from the other end.

D–C

Questions

(Grades G-E)

1 Complete this equation:
momentum = mass × ....

(Grades D-C)

2 How does momentum differ from kinetic energy?

(Grades G-E)

3 Two inflated balloons, one small and one large, are resting on a table. What is the momentum of the balloons?

(Grades D-C)

4 One of the balloon bursts and shoots off the table. What can you say about the momentum of the moving balloon and the air that bursts out of it?

Keep it safe

Force causes a change in momentum

- **Momentum** is conserved as long as *no* **force** acts on an object.
- A force *changes* an object's momentum by changing its **velocity** or **direction**.
- The change in momentum depends on the **size** of the force and the **time** it acts.
- The diagram opposite shows how a small braking force changes the car's momentum slowly. A large force from the tree changes the momentum quickly.

Different ways of changing the car's momentum.

G–I

Changing momentum safely

- In a head-on collision, a car's seat belts let passengers move forward slightly before stopping them. They change momentum more slowly than the car, avoiding injury.
- On impact, the car bonnet crumples. This: slows the *rate of change* of the car's momentum; absorbs some of the car's **kinetic energy**; and transfers less energy to passengers.

D–C

Static electricity

Static electricity

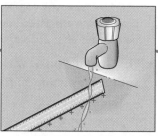

- Rubbing some materials together builds up **static electricity** on them.
- This happens because one material loses **electrons** and the other gains them. Electrons are **negative electrically** charged particles.
- Similar (like) charges **repel**. Opposite (unlike) charges **attract**.
- The charged ruler attracts the fine stream of water, because the water contains charged particles.

The charged plastic ruler attracts the stream of water, which curves towards the ruler.

G–E

Investigating static electrical charge

- In an atom, **protons** carry **positive charge** and **electrons** carry **negative charge**.
- An **electrical insulator** is a material that will hold static electrical charge and not let it flow away.
- When two insulators are rubbed together:
  - one gains electrons (e.g. polythene), becoming negatively charged
  - the other loses electrons (e.g. cellulose acetate) and is left with an equal positive charge.
- The suspended strip of cellulose acetate in the diagram carries a positive charge. If the strip approaching it:
  - is also charged cellulose acetate, the like charges repel
  - is charged polythene, the strips attract because they carry unlike charges.

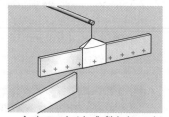

A charged strip (left) is brought close to a positively charged cellulose acetate strip.

D–C

Questions

Grades G-E

1 An ice hockey puck slides across the ice. It hits a player's foot and moves off in a different direction, at the same speed. Has its momentum changed? Explain your answer.

Grades D-C

2 Use the idea of momentum to explain how wearing a seat belt helps to avoid head injuries in a car crash.

Grades G-E

3 A camper's shirt sleeve rubs against the fabric of her tent. The tent fabric becomes negatively charged. What is the name for this kind of electric charge?

Grades D-C

4 What has moved, and in which direction, to cause the tent fabric to become negatively charged?

Charge

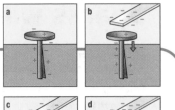

Charge and force

- **Electrons** give an object a negative **static electrical charge**.
- Removing electrons gives an object a positive charge.
- Because charges exert a **force** on each other:
  - like charges (both positive or both negative) **repel**
  - unlike charges (positive and negative) **attract**.
- **Insulators** (e.g. plastics) hold electrical charge.
- **Electrical conductors** (e.g. metals, water) allow charge (electrons) to flow as an **electrical current**.
- A **gold leaf electroscope** detects if an object has an electrical charge. The gold leaf and rod are insulated.

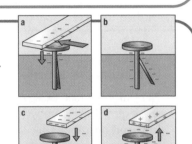

a The gold leaf electroscope is **uncharged**. It has equal numbers of **positive** and **negative** charges all over it.
b The negative electrical charge on the strip repels the negative electrical charge on the electroscope. They move as far away as they can.
c The metal rod and the gold leaf both get a negative electrical charge. They move away from each other. The gold leaf rises.
d The more electrical charge the strip has, the more the gold leaf rises.

Identifying charges

- The electroscope can also tell whether a charge is positive or negative.

a Wipe a negatively charged polythene strip across the cap of the electroscope. The negative charges try to get as far apart as possible, so some of them will move down onto the electroscope.
b Remove the polythene strip. The electroscope now has a negative electrical charge. The gold leaf stays up.
c Holding a negatively charged object near the electroscope repels more negative charges down the metal rod of the electroscope. The gold leaf rises more.
d Holding a positively charged object near the electroscope attracts negative charges up to the cap. The gold leaf falls.

Van de Graaff generator

The Van de Graaff generator and lightning conductor

- In a **Van de Graaff generator**, two materials rub together and build up equal and opposite static electrical charges.
- The dome gains a large but safe **positive charge**. The **potential difference (voltage)** is high.
- When a girl standing on an insulating mat touches the dome, the positive charge redistributes over her and the dome. Her positively charged hairs repel each other. They move apart, standing on end.

Positive charge causes hairs to repel each other.

Moving charges

- The cloud's negative static charge induces a positive charge on the earthed lightning conductor. This charge streams up towards the cloud, discharging it and reducing the chance of a lightning strike.

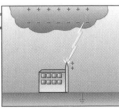

The earthed conductor discharges the cloud.

Questions

(Grades G-E)
1 Why cannot static electricity build up on an electrical conductor that is earthed?

(Grades D-C)
2 Look at the diagrams of the electroscope in the second panel. Explain why the gold leaf goes up after the polythene strip is wiped across the cap.

(Grades G-E)
3 Why do the hairs on the girl's head repel each other, when she touches the dome of the Van de Graaff generator?

(Grades D-C)
4 What happens if you connect a charged object to earth with a conductor?

Sparks will fly!

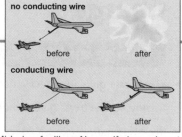

Danger of electrostatic charge

- **Friction** builds up **static electrical (electrostatic) charge** on **insulating** materials.

- Charge can jump as a **spark** (flow of charge) between insulated objects, and set fire to flammable materials.

- **Earthing** prevents build-up of charge. Where this is not possible, charge can be redistributed evenly so that charge does not flow.

- In an operating theatre, fluids and gases are flammable. Plastics can become charged. They are **discharged** safely by keeping air humid, so that it conducts electricity, and installing a floor of material that conducts (earths) charge.

- Aircraft can build up static charge as they move through the air. Before mid-air refuelling, a wire connects the two aircraft to redistribute charge evenly between them.

- Friction in high pressure hoses charges the water droplets. Hoses used to clean oil tankers are earthed to prevent sparks from igniting the oil vapour.

Mid-air refuelling. Above, if charge is not evenly distributed, a spark may explode the fuel. Below, distributing charge evenly prevents sparks.

Making use of static electricity

- Paint droplets gain positive charge as they stream through the spray gun. The car bodywork is earthed, so a negative charge is **induced** near the spray. It attracts the paint droplets so they are deposited evenly over the bodywork.

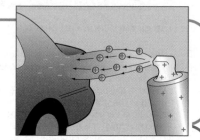

Positively charged paint droplets are attracted to the car's bodywork.

- Flue gases from power stations must be cleaned to reduce pollution. **High voltage** wires in the chimney are negatively charged. Air particles are **ionised** between the wires. Positively charged particles stay near the wires. Negative charges are picked up by **ash** and **dust** particles which move to earthed metal plates lining the chimney. There they are collected.

- The drum inside a **photocopier** is positively charged. When you press Copy, light reflected from white areas of a sheet being photocopied illuminates the drum. Charge from those areas flows away, leaving positive charge only where there is colour on the sheet. These positively charged areas of the drum attract particles of black toner, which is then transferred to the copy.

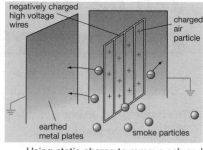

Using static charge to remove ash and dust from flue gas in power stations.

How Science Works

You should be able to:
- explain why static electricity is dangerous in some situations and how precautions can be taken to ensure that the electrostatic charge is discharged safely
- explain how static electricity could be useful.

Questions

(Grades G-E)

1 Why could static electricity cause an explosion during mid-air refuelling of an aircraft?

(Grades D-C)

2 Explain why the high-pressure water hoses that are used to clean oil tanks are always made of metal and not plastic.

(Grades G-E)

3 How does static electricity help to stop paint being wasted when a car is sprayed?

(Grades D-C)

4 The toner in a photocopier has a negative charge. Why do particles of toner stick to some parts of the drum but not others?

P2a summary

Distance-time graphs help us to 'picture' how an object moves:
- the **slope** of the graph shows the **speed**.

$$\text{average speed} = \frac{\text{total distance travelled}}{\text{total time travelled}}$$

The **velocity** of an object tells us its **speed** in a given **direction**. **Acceleration** is the **rate of change** of velocity.

$$\text{acceleration} \ (\text{m/s}^2) = \frac{\text{change in velocity (m/s)}}{\text{time taken for change (s)}}$$

Velocity-time graphs show how the velocity changes:
- the **slope** shows the acceleration
- the **area** under the graph shows the **distance travelled**.

Forces can add up or cancel out to give a **resultant force**.

When the resultant force is *not* zero, an object accelerates.

The **acceleration** is:
- directly proportional to **force**, $a \propto F$
- inversely proportional to **mass**, $a \propto 1/m$

$$\text{resultant force (N)} = \text{mass (kg)} \times \text{acceleration (m/s}^2)$$
$$F = ma$$

For every force there is an equal and opposite force.

Balanced forces have the same value but act in opposite directions, e.g. a person floating in water has balanced forces.

Objects moving at **constant velocity** (including **terminal velocity**) have balanced forces.

$$\text{stopping distance} = \text{thinking distance} + \text{braking distance}$$
$$\text{thinking distance} = \text{speed} \times \text{reaction time}$$

Motion and forces

Stopping distance increases with:
- increasing **speed**
- a greater **mass**
- reduced **friction**
- increased **reaction time**.

Air resistance (in gases) and **drag** (in liquids), from frictional forces between the fluid and an object, increase as:
- the object's **speed** increases
- the fluid gets **denser**.

An object **falling** through a **fluid**, **accelerates** until it reaches **terminal velocity**, when the **resultant force** is **zero**:

upthrust (fluid's resistant force) = object's weight due to gravity

Every **moving** object has **kinetic energy** that can be transformed into other forms.

An object's kinetic energy increases when its:
- **mass** increases
- **speed** increases.

Using a **force** to do **work** on an object gives it **energy**:

work done = energy transferred

The amount of work done depends on **force** and **distance**:

work done (J) = force (N) × distance moved in the direction of the force (m)

Every **moving** object has **momentum** that depends on its **mass** and its **velocity**:

momentum (kg m/s) = mass (kg) × velocity (m/s)

Momentum has **size** and **direction** and is **conserved** in collisions and explosions.

Forces change an object's momentum by changing its **velocity** or **direction**. The bigger the force on a moving object, the larger its **rate of change** of momentum.

Rubbing **electrical insulators** together can build up **static electricity**, because **electrons** transfer from one material to another.

If an object **gains** electrons, it has a **negative** charge. If it **loses** electrons, it has a **positive** charge.

Similar (like) charges **repel**. Opposite charges **attract**.

Static electricity

A **gold leaf electroscope** shows the **size** and **type** of charge on an object.

Static electricity can be dangerous if **sparks** cause flammable materials to **ignite** or **explode**. **Earthing** prevents a build-up of charge.

Paint sprayers, smoke precipitators and photocopiers are all useful **applications** of static electricity.

Circuit diagrams

Standard symbols and circuit diagrams

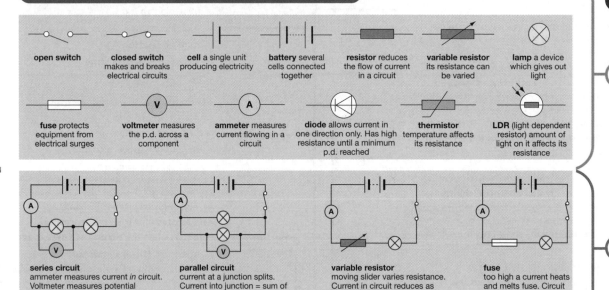

open switch

closed switch makes and breaks electrical circuits

cell a single unit producing electricity

battery several cells connected together

resistor reduces the flow of current in a circuit

variable resistor its resistance can be varied

lamp a device which gives out light

fuse protects equipment from electrical surges

voltmeter measures the p.d. across a component

ammeter measures current flowing in a circuit

diode allows current in one direction only. Has high resistance until a minimum p.d. reached

thermistor temperature affects its resistance

LDR (light dependent resistor) amount of light on it affects its resistance

series circuit ammeter measures current *in* circuit. Voltmeter measures potential difference (p.d.) *across* lamp. Current is *same* at any point in circuit

parallel circuit current at a junction splits. Current into junction = sum of currents in splits

variable resistor moving slider varies resistance. Current in circuit reduces as resistance increases, and increases as resistance reduces

fuse too high a current heats and melts fuse. Circuit breaks, protecting lamp

G–E

D–C

Resistance 1

How Science Works

You should be able to: interpret and draw circuit diagrams using standard symbols.

Electrons as energy carriers

- **Electrical current** is the flow of **electrons** carrying negative **charge**.

- **Free electrons** move between the atoms of a conductor.

- **Collisions** with atoms slow the free electrons down. This is **resistance**. The more collisions, the greater the resistance and the smaller the current.

- Copper has very low resistance, so allows for large currents.

copper wire

e^- electron

copper atoms

Electrons travel through the spaces between the atoms of the copper wire.

G–E

Resistance

- A **potential difference** (**voltage**) exerts a **force** that pushes free electrons along in a circuit.

- Since all electrons have a like (negative) charge, electrons fixed in atoms repel free electrons. The **force of repulsion** slows down moving electrons. This causes resistance.

- Free electrons in **metals** make them good conductors with low resistances.

- **Insulators** do not have free electrons and so are poor conductors with very high resistances.

Top Tip!

Potential difference is the amount of 'push' a battery has to move current around a circuit. Unit of potential difference = volt (V) Symbol for potential difference = V

D–C

Questions

(Grades G-E)

1 Without looking, draw the symbols for a lamp and a resistor.

(Grades D-C)

2 Draw a series circuit containing two ammeters, one lamp and a voltmeter measuring the potential difference across the lamp.

(Grades G-E)

3 What are the particles that carry charge in an electrical circuit?

(Grades D-C)

4 Why are metals good conductors?

Resistance 2

Factors affecting resistance

- Long wires have greater **resistance** than short wires. Thin wires have greater resistance than thick wires.

- Electrical resistance is due to electrons colliding with atoms. Electrons in a longer wire have more collisions. Increasing **length** *increases* the wire's resistance.

- A thick wire has more electrons free to flow as a **current** than a thin wire. Increasing **thickness** *decreases* the wire's resistance.

- As **temperature** rises, atoms in a wire gain **kinetic energy** and **vibrate** more vigorously. They **repel** and **deflect** electrons more frequently and with greater energy. **Heating** *increases* a wire's resistance.

- **Different materials** have different resistances, depending on the size of atoms and the distance between them. High resistance can cause energy to be transformed into light and heat, as in filament light bulbs and electric bar heaters.

Ohm's Law

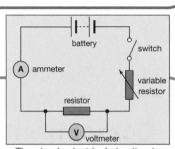
The simple electrical circuit set up by Georg Ohm.

Ohm and resistance

- **Georg Ohm** investigated how **potential difference** relates to **current**, using the circuit shown. As he adjusted the variable resistor, he recorded potential difference across the **resistor** and the current through it.

- Ohm's straight-line graph shows that:
  – resistance of the **resistor** is *constant*

  $$\text{resistance} = \frac{\textbf{potential difference}}{\textbf{current}} = \textbf{Ohm's Law}$$

  – current is *directly proportional* to potential difference across the resistor.

- The equation shows that, for a given potential difference across a component, the greater the resistance, the smaller the current.

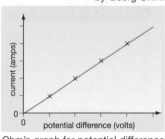
Ohm's graph for potential difference plotted against current.

Resistance of components

- To find the resistance of a **component**:
  – the component replaces the resistor in the circuit diagram above
  – current through it and potential difference across it are measured.

- The graph for a **filament lamp** shows that current levels off as potential difference increases. This is because resistance increases with heating.

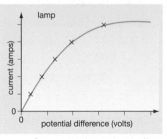
Current begins to level off as potential difference increases.

Questions

(Grades G-E)

1 Two series circuits are identical, except that circuit A has long, thin wires connecting the components, and circuit B has short, thick wires. In which circuit is resistance greater?

(Grades D-C)

2 When a current flows through the metal wire in a filament lamp, it gets hot. How does this affect its resistance?

(Grades G-E)

3 Write down the equation that links current, potential difference and resistance (Ohm's Law).

(Grades D-C)

4 Why does a filament lamp not obey Ohm's Law?

More components

Thermistors, LDRs and how they work

- When the **temperature** of a **thermistor** is increased, its resistance *decreases*.

- So, as the thermistor heats up, the current through it *increases*.

- A thermistor is a semiconductor material (e.g. germanium, silicon). Heating frees more electrons, so increasing the current.

- Thermistors are used in equipment for detecting temperature changes, including electronic thermometers and fire alarms.

- When **light** is shone on a **light dependent resistor (LDR)**, its resistance *decreases*, so more current flows.

- As the light intensity increases, the current through the LDR *increases*.

- An LDR is made from the semiconductor compound **cadmium sulfide**. Light transfers energy that frees more electrons to carry current.

- LDRs are used in smoke detectors, automatic light controls and burglar alarms.

How Science Works

You should be able to: apply the principles of basic electrical circuits to practical situations.

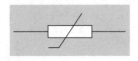

Symbol for a thermistor and a bead thermistor.

An LDR and its symbol.

Components in series

Effect on current of adding lamps to a series circuit

- Components in a **series circuit** are connected in one *continuous* loop.

- The diagram shows lamps 1, 2 and 3 added to a circuit one at a time.

- The brightness of the lamps gets less with each lamp added.

- At any time, the current is the *same* everywhere in the circuit.

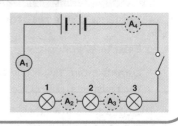

Three lamps connected in series.

Series circuits, resistance and batteries

- In the series circuit, voltmeters V_1, V_2 and V_3 measure potential difference (p.d.) across each lamp. V_4 measures p.d. across all three lamps.
  - p.d. measured by V_4 = sum of p.d.s across lamps = p.d. of battery
  - i.e. total p.d. is shared between all components

- Ohm's Law, resistance = p.d./current, gives the resistance of each lamp from p.d. and current measurements.
  - In a series circuit, total resistance = sum of resistances of components.

- A **battery** is several **cells** connected in series.

- With all cells connected plus to minus, p.d. of battery = sum of p.d.s of cells.

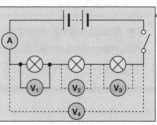

Measuring p.d. across three lamps connected in series.

Top Tip!

Remember: In a series circuit, total p.d. is the *sum* of the p.d.s of each cell. This p.d. is *shared* between the components.

Questions

Grades G-E
1 What happens to the resistance of an LDR as light intensity increases?

Grades D-C
2 Give **one** use of an LDR.

Grades G-E
3 What happens to the brightness of a lamp in a series circuit, as more lamps are added?

Grades D-C
4 A cell has a p.d. of 1.5 V. What is the p.d. across four of these cells connected in series?

Components in parallel

Current and potential difference in parallel circuits

- At **junctions** (**splits**) in a **parallel circuit** (see diagram), the current divides to take different routes, then joins up again.

- Ammeters A_1 and A_5 measure the same **total current**.

- As each lamp is connected, the **brightness** remains the *same*.

- Ammeters A_2, A_3 and A_4 measuring current through each lamp show that:

$$A_1 = A_5 = A_2 + A_3 + A_4$$

- For parallel circuits, the *total current* through the whole circuit is the *sum* of the currents through the separate components.

- Voltmeters (see diagram), show that there is the same **potential difference** across the battery and each lamp:

$$V_1 = V_2 = V_3 = V_4$$

- For parallel circuits, the *potential difference* across the battery and each component is the *same*.

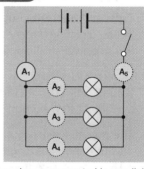

Three lamps connected in parallel.

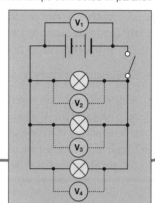

Measuring potential difference.

The three-pin plug

Structure and wiring of a three-pin plug

- The photograph shows the structure and wiring of a **three-pin plug**.

- The plastic covering the wires is colour coded:
  - **yellow** and **green** striped wire is **earth**, connected to top pin
  - **blue** wire is **neutral**, connected to left pin
  - **brown** wire is **live**, connected through fuse to right pin
  - **fuse** (see page 126) joins brown wire and right pin.

Top Tip!

Which goes where?
b**R**own = right
b**L**ue = left

Wiring up a three-pin plug.

Cables and plugs

- A **cable** and a three-pin plug connect most electrical appliances to the mains. **Insulating** material encloses all but the pins.

- The cable carries the three wires, live, neutral and earth, or '**twin** and **earth**'. **Insulating** plastic coats each wire. A plastic sheath encloses all three.

- Strands of each wire should be secured under their connector.

- **Twin-core cable**, with live and neutral wires only, is used in **double insulated** appliances such as a hairdryer and an electric drill.

- When the **fuse** blows, it isolates the live wire and no current flows.

- The **cable grip** prevents wires from being pulled from the plug by accident.

How Science Works

You should be able to: recognise errors in the wiring of a three-pin plug.

Questions

Grades G-E
1 In the first circuit diagram, the current flowing through ammeter A_1 is 12 A. What is the current measured by ammeter A_5?

Grades D-C
2 In the second circuit diagram, the cells provide a potential difference of 4.5 V. The three lamps are identical. What is the p.d. across each lamp?

Grades G-E
3 What is the colour of the earth wire in a three-pin plug?

Grades D-C
4 Why does a plug contain a fuse?

Domestic electricity

Direct current and alternating current

- **Cells** and **batteries** supply **electricity** as **direct current**, **d.c.**, in one direction. Potential difference is low, typically 1.5 V or 9 V, to power small appliances (e.g. torches, cameras).

- Domestic electricity has **alternating current**, **a.c.** It constantly changes direction. Potential difference averages 230 V, enough to power hairdryers and kettles.

- The potential difference of a.c. changes (oscillates) from plus to minus (see diagram above) at 50 times per second, 50 Hz – its **frequency**.

- The **cathode ray oscilloscope** (**CRO**) displays **potential difference** measured in volts (see diagram on the right). The x-axis represents time.

- Direct current gives a straight-line, positive trace.

- As the diagram on the right shows, alternating current gives an S-shaped trace. The x-axis is at zero potential difference, values above it are positive and those below are negative. The maximum height of the wave gives the **peak** potential difference.

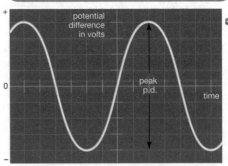

Oscilloscope trace showing the voltage of alternating current (left) and direct current (right) against time.

> **Top Tip!**
>
> Mains voltage is often written as: (220–250) V, which means that it can range between this lower and upper limit; 230 V is accepted as an average value.

An oscilloscope trace for alternating current.

> **How Science Works**
>
> You should be able to: compare potential differences of d.c. supplies and the peak potential differences of a.c. supplies from diagrams of oscilloscope traces.

Safety at home

Fuses and earth wires

- A **fuse** is connected in series in a circuit. If a current is too high, the fuse wire heats and melts, breaking the circuit.

- Fuses have different ratings, appropriate to the current drawn by appliances.

- Appliances with metal casings have an **earth wire**. If the live wire touches the metal casing, a large current from the live wire to the earth wire melts the fuse, breaking the circuit.

- When electrical faults occur, fuses and earth wires protect appliances from damage and people from injury or death.

> **Top Tip!**
>
> The fuse and earth wire together protect the appliance and the user.

Circuit breakers

- Modern fuse boxes use **miniature circuit breakers** (**MCB**). An electromagnet or electronic mechanism immediately breaks the circuit when there is a tiny increase in current.

- A **residual current device** (**RCD**) breaks a circuit in less than 0.05 s after detecting a difference in the current in live and neutral wires. It makes lawn mowers and hedge trimmers safe to use.

A residual current device is essential wherever a cable might be cut by accident.

Questions

(Grades G-E)
1 What is a.c. short for?

(Grades D-C)
2 What does the trace of direct current look like on an oscilloscope screen?

(Grades G-E)
3 A fuse has a rating of 0.5 A. What size of current will melt the fuse?

(Grades D-C)
4 Suggest why having an RCD in a circuit is even safer than a fuse.

Which fuse?

Calculating the correct fuse rating

You should be able to: calculate the current through an appliance from its power and the potential difference of the supply, and from this determine the size of the fuse needed.

G–E

- Every appliance has a **power rating**. Power (see page 55) is measured in **watts** (**W**). Its correct **fuse** depends on the current it takes from the mains (p.d. = 230V).

  power = **p.d.** × **current**

  so **current** = **power/p.d.**

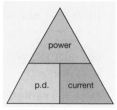

The power triangle.

- An appliance has a power rating of 1000W.

  So current = 1000 W ÷ 230V = 4.3A

  - A 3A fuse would melt when the appliance is switched on.
  - A 13A fuse would not melt if a fault occurred.
  - The correct fuse is 5A.

D–C

- In a resistor (e.g. electric fire), electrical energy is transformed into heat.

- The **rate** at which energy is transformed is called **power**:

  $$\text{power (watts)} = \frac{\text{energy transformed (joules)}}{\text{time taken (seconds)}}$$

  - A 100 W light bulb transforms electrical energy to light and heat at a rate of 100 J/s.

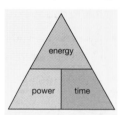

The energy/power triangle.

Radioactivity

particle	relative mass	charge
proton	1	+1
neutron	1	0
electron	0	–1

Atoms and atomic particles

G–E

- An atom has a nucleus made up of **protons** and **neutrons**, surrounded by **electrons**.

- The table shows the characteristics of these atomic particles.

- Atoms have no net charge: **number of protons = number of electrons**.

- An atom that loses an electron (–1) becomes an **ion** with a charge of +1.

- An atom that gains an electron becomes an ion with a charge of –1.

- All atoms of a particular element have the *same* number of protons. Atoms of different elements have *different* numbers of protons.

D–C

- The number of:
  - protons in an atom is its **atomic number**
  - protons + neutrons is its **mass number**.

- The symbol for lithium can be written as shown:

mass number = protons + neutrons

atomic number = protons

$^{7}_{3}\text{Li}$

Isotopes

D–C

- Atoms of a particular element may have different numbers of neutrons. Such atoms have the same atomic number but different mass numbers.

- These different forms of an element are **isotopes**.

- The nuclei of some isotopes are unstable. They emit energy as particles or rays called **radiation**.

Questions

(Grades G-E)

1 What is the unit in which power is measured?

(Grades D-C)

2 A hairdryer has a power rating of 1500W. If the mains supply is 230V, what current flows through the hairdryer?

(Grades G-E)

3 Which kind of particle in an atom has a charge of +1?

(Grades D-C)

4 This is the full symbol for boron: $^{11}_{5}\text{B}$. How many protons does it have in its nucleus? How many neutrons does it have in its nucleus? What is its mass number?

Alpha, beta and gamma rays 2

Types of radiation

- In some **isotopes**, the *balance* of protons and neutrons in the nucleus is **unstable**.
  - To reach a stable state the nucleus emits **radiation**, which carries energy.

- Radiation can knock electrons from atoms. They become **ionised**.

- **Alpha (α) radiation** is a particle containing 2 protons and 2 neutrons – a helium nucleus. It is large and slow.

- Being large, heavy and carrying a +2 charge, it readily ionises surface atoms.

- **Beta (β) radiation** is a particle. It is a very small, fast, high energy electron.

- It is negatively charged, e⁻. Being small, it tends to miss atoms.

- **Gamma (γ) radiation** is an electromagnetic wave with no mass or charge, moving at the speed of light and carrying high energy.

- With no mass or charge, it passes between the atoms of most materials.

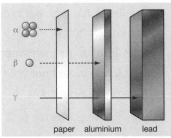

The ability of different radiations to penetrate different materials depends on their energy and mass (size).

Background radiation 2

Sources of background radiation

- **Background radiation** comes from the environment all the time.

- Generally, radiation is at a low level that is harmless to living things.

- Its sources include radioactive materials in rocks and soil, and cosmic rays from space. Some background radiation is due to the use or manufacture of radioactive materials by humans.

- Background radiation is detected by a **Geiger counter**, heard as intermittent clicks.

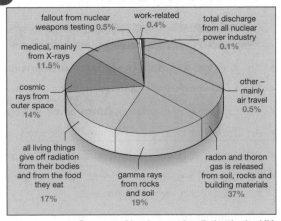

Sources of background radiation in the UK.

Sources that add to background radiation

- **Granite** rock contains radioactive materials that decay to give **radon** gas, itself radioactive and emitting harmful radiation.

- The Earth's magnetic field and the atmosphere shield the ground from most harmful **cosmic rays** from the Sun. This shielding is weaker above the Earth, so pilots and air crew are more exposed.

- Medical staff using X-rays, and others working with radioactive materials, risk higher radiation levels. They wear a radiation badge to monitor their exposure.

Questions

Grades G-E
1 What is meant by a 'radioactive isotope'?

Grades D-C
2 State the charges on: **a** an alpha particle; **b** a beta particle; **c** gamma radiation.

Grades G-E
3 How can we detect background radiation?

Grades D-C
4 What is the major source of background radiation in the UK? How can this explain the high levels of background radiation in places such as Cornwall, where there is a lot of granite?

Inside the atom

Experiments into the structure of the atom

- Alpha particles have a +2 charge. To investigate atomic structure, Rutherford and Marsden fired alpha particles at a sheet of very fine gold foil. They found that:
  - most went through, either straight or at angles
  - a few bounced back at different angles. Something repelled them
  - very few came straight back.

- They concluded that something in gold atoms was repelling the alpha particles.

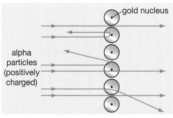

The scientists' model to explain what happened in their investigation.

- Their results suggested that:
  - most of an atom is empty space
  - it has a very small, dense, positive nucleus
  - electrons are not part of the nucleus, but orbit round it.

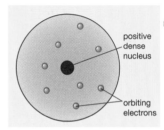

Rutherford's model of the atom.

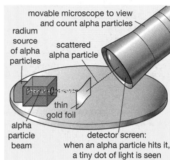

Model of scattering apparatus.

Nuclear fission

Splitting the atom

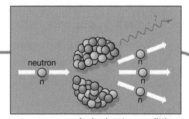

A single atom splitting.

- When an atomic nucleus is split in **nuclear fission**, some of its mass is converted to energy.

- Neutrons are fired at very large **unstable isotopes**. The nucleus of an unstable atom absorbs a neutron, splits in two and releases **nuclear energy**. It also produces more neutrons, setting up a **chain reaction**.

- Uranium-235 and plutonium-239 have unstable nuclei and so are **fissionable** substances. They are used to produce nuclear energy.

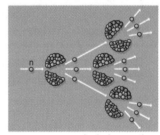

A chain reaction.

- An unstable isotope such as uranium-235 does not normally produce energy. A sample has to be pure uranium, and at least tennis ball-sized, so that neutrons do not leave the mass without setting up a chain reaction. This minimum amount is the **critical mass**.

- The nuclear bomb is an example of an uncontrolled chain reaction.

- Nuclear power stations control chain reactions very carefully.

How Science Works

You should be able to: sketch a labelled diagram to illustrate how a chain reaction may occur.

Questions

Grades G-E

1 Name the **two** scientists whose experiments with gold foil and alpha particles helped to find out the structure of the atom.

Grades D-C

2 How did their results suggest that atoms had a very small nucleus surrounded by empty space?

Grades G-E

3 What does 'nuclear fission' mean?

Grades D-C

4 What must a uranium-235 atom absorb, to make it undergo fission?

Nuclear power station

The nuclear reactor

- Nuclear power produces a huge amount of energy as heat from a small amount of **nuclear fuel**, such as uranium-235.

- The energy heats water under high pressure. The pressurised hot water passes through a **heat exchanger** where a separate water supply is boiled to super-heated steam. This turns a turbine connected to a generator that produces **electricity**.

- To maintain a safe, controlled nuclear reaction:
  - rods of graphite, a **moderator** material, slow down neutrons so that they will be absorbed by fuel nuclei; water also acts as a coolant
  - movable **control rods** of boron or cadmium absorb excess neutrons.

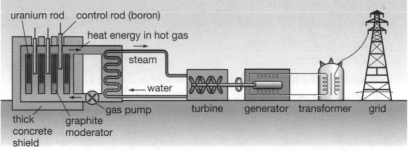

In a nuclear power station, energy from nuclear fission generates steam.

Nuclear fusion

The birth of a star

- A star is formed over millions of years when dust and smaller particles in a **nebula** clump together, attracted by gravity. (See also page 66.)

- The mass of material heats up. At 15 million Kelvin, there is enough energy to fuse hydrogen nuclei together and form helium nuclei. This is **nuclear fusion**:

A nebula in which stars are born.

nuclear fusion

2 hydrogen nuclei ⟶ 1 helium nucleus + ENERGY

Proton-proton reactions

- A hydrogen nucleus is a **proton**. Protons carry positive charge.

- It takes an immensely high temperature to *start* a **fusion reaction**. Protons need sufficient **kinetic energy** to move at very high speeds and overcome the force of repulsion between their like charges. Once started, the reaction itself produces great energy.

- As a result, the **reaction cycle** continues, generating even more energy.

- Similar reactions occur in stars and supernovae (see page 67).

Top Tip!

In nuclear fission, the nucleus of a very large atom splits. In nuclear fusion, the nuclei of two small atoms join together.
Both processes produce huge amounts of energy.

Questions

Grades G-E
1 Give **one** example of a nuclear fuel.

Grades G-E
2 Explain the difference between 'nuclear fission' and 'nuclear fusion'.

Grades G-E
3 How is energy released inside a star?

Grades G-E
4 Name the new element that is formed when the nuclei of two hydrogen atoms fuse together.

P2b summary

Electrical current is the flow of **electrons** carrying negative **charge**. Electrical energy is transformed to heat energy when electrical charge flows through a **resistor**.

Electrical **resistance** in wires is due to **free electrons** colliding with atoms.

Resistance is *increased* in:
- **long** wires
- **thin** wires
- **heated** wires.

Ohm's Law states that:
$$\text{resistance} = \frac{\text{potential difference}}{\text{current}}$$

In the **UK**, domestic electricity has an **alternating current**, 230 V, a **frequency** of 50 Hz and is extremely dangerous.

An **oscilloscope** shows the **potential difference** (voltage) and **frequency** of electricity.

The **rate** at which energy is **transformed** is called **power**:
$$\text{power (W)} = \frac{\text{energy transformed (J)}}{\text{time taken (s)}}$$

Isotopes are the different forms of an **element** with the same number of **protons** but a different number of **neutrons** in their atoms.
Isotopes of an element have the same **atomic number** but different **mass numbers**.

Isotopes with **unstable nuclei** emit **energy** as particles or rays called **radiation**.

An **alpha particle** is a helium ion. A **beta particle** is a high energy electron.
A **gamma ray** is an electromagnetic wave with no charge or mass.

Nuclear fission is the **splitting** of an atomic **nucleus**.
Neutrons are fired at unstable isotopes and **energy** is released. This produces more neutrons and sets up a **chain reaction**.

Resistance

The resistance of a **light-dependent resistor** (LDR) decreases as light intensity increases.

The resistance of a **thermistor** decreases as the temperature increases.

The resistance of a **filament lamp** increases as the temperature of the filament increases.

Insulators do not have free electrons and so are **poor conductors** with very **high resistance**.

Components can be connected in **series** and in **parallel**.

Domestic electricity

Fuses and **earth wires** protect appliances from damage and people from injury or death. **Three-pin plugs** must be correctly wired and hold the correct fuse for the appliance.

The **current** through an appliance and the **size of the fuse** required can be calculated by:
power = current × potential difference

Radioactive substances

Background radiation comes mainly from rocks, soil, cosmic rays, living things and medical X-rays.

Rutherford and Marsden revealed the **structure** of the **atom** using alpha particle scattering experiments.

An **atom** consists of:
- a small dense **nucleus** of **protons** and **neutrons**
- fast-moving **electrons** orbiting round the nucleus
- lots of empty **space**.

Nuclear fusion and nuclear fission

Uranium-235 and **plutonium-239** are fissionable elements and are used to produce **nuclear energy** in nuclear power stations.

Nuclear fusion is the **joining** of two smaller **nuclei** to form a larger one.
Stars release energy by nuclear fusion:
2 hydrogen nuclei → 1 helium nucleus + ENERGY

131

Checklists – Science Unit 1

B1a Human biology

I know:

how human bodies respond to changes inside them and to their environment

- [ ] nerves and hormones coordinate body activities and help control water, ion and blood sugar levels and temperature
- [ ] receptors detect stimuli; information passes to the brain; and effectors (muscles and glands) respond
- [ ] reflex actions are automatic and very fast; they involve sensory, relay and motor neurones, with synapses between them, in a reflex arc
- [ ] the menstrual cycle is controlled by FSH, LH and oestrogen

what we can do to keep our bodies healthy

- [ ] a balanced diet contains the right balance of all the nutrients and energy humans need, and regular exercise helps you to stay fit
- [ ] different people have different energy needs
- [ ] obesity increases the risk of arthritis, diabetes, high blood pressure and heart disease
- [ ] too much cholesterol in the form of LDL in the blood increases the risk of heart disease

how we use/abuse medical and recreational drugs

- [ ] drugs affect people's behaviour, can damage the brain and some hard drugs are addictive
- [ ] some drugs are legal (alcohol) and some are illegal (cocaine, heroin)
- [ ] the substances in tobacco smoke cause many diseases, e.g. cancer and bronchitis
- [ ] drugs can be beneficial, but must be thoroughly tested before use

what causes infectious diseases and how our bodies defend themselves against them

- [ ] pathogenic microorganisms cause infectious diseases
- [ ] some white blood cells (phagocytes) ingest pathogens and kill them; and others (lymphocytes) make antibodies
- [ ] antibiotics kill bacteria inside the body but do not kill viruses; however, some bacteria can develop resistance to antibiotics
- [ ] immunisations and vaccinations offer protection from various diseases

B1b Evolution and environment

I know:

what determines where particular species live

- [ ] animals and plants are adapted to live in different habitats, to compete for resources and to survive attack from predators
- [ ] caribou and camels are adapted to extreme environments

why individuals of the same species are different from each other; and what new methods there are for producing plants and animals with desirable characteristics

- [ ] DNA is the genetic material that controls inherited characteristics
- [ ] reproduction can be sexual (variation in offspring) or asexual (identical offspring)
- [ ] cloning and genetic engineering can be used to produce plants and animals with desirable characteristics

why some species of plants and animals have died out and how new species of plants and animals develop

- [ ] theories of evolution have changed over time; Darwin proposed the theory of evolution by natural selection
- [ ] fossils tell us how present-day species have evolved and how they compare to prehistoric species
- [ ] species can become extinct due to environmental change and human impact

how humans affect the environment

- [ ] increases in human population use up more resources and produce more waste and pollution
- [ ] human action contributes to acid rain, air pollution, water pollution, and over-use of land
- [ ] living organisms, such as lichens, can be used as indicators of pollution
- [ ] increasing the greenhouse effect can lead to global warming
- [ ] sustainable development, e.g. using renewable energy resources and recycling, can help to safeguard the environment for future generations

C1a Products from rocks

I know:

how rocks provide building materials

- [ ] limestone, calcium carbonate, can be used to make cement, concrete and glass which are used as building materials
- [ ] an element consists of one type of atom; two atoms of the same element can join together to form a molecule; a compound consists of two or more elements joined together
- [ ] atoms are held together in molecules and lattices by chemical bonds, which involve giving, taking or sharing electrons

how rocks provide metals and how metals are used

- [ ] metals are extracted from their ores, often oxides, by reduction with carbon (iron), electrolysis (aluminium and copper) or other chemical reactions (titanium)
- [ ] metals are mixed together to make alloys (e.g. iron and other metals or carbon make steel)

- [ ] aluminium is expensive to produce, is often too soft on its own, but forms strong alloys with other metals
- [ ] copper is a hard, strong, good conductor and can be used for wiring and plumbing
- [ ] aluminium and titanium are resistant to corrosion and have low density

how we get fuels from crude oil

- [ ] crude oil is a mixture of hydrocarbon compounds that can be separated by fractional distillation. Some of the fractions can be used as fuels
- [ ] most of the compounds in crude oil are saturated hydrocarbons called alkanes, which have the general formula C_nH_{2n+2}
- [ ] the burning of fossil fuels releases useful energy but also harmful substances, e.g. sulfur dioxide causes acid rain; carbon dioxide causes global warming; smoke particles cause global dimming

C1b Oils, Earth and atmosphere

I know:

how polymers and ethanol are made from oil

- [ ] crude oil is made from hydrocarbons that can be cracked to form alkenes, which are unsaturated hydrocarbons and have the general formula C_nH_{2n}
- [ ] alkenes can be made into polymers, which are long chain molecules created when lots of small molecules called monomers are joined together in polymerisation
- [ ] polymers can be used to make useful substances e.g. waterproof materials and plastics, but many are not biodegradable

how plant oils can be used

- [ ] vegetable oils can be hardened to make margarine in a process called hydrogenation
- [ ] biodiesel fuel can be produced from vegetable oils
- [ ] oils do not dissolve in water. They can be used to produce emulsions e.g. in salad dressings

what the changes are in the Earth and its atmosphere

- [ ] the Earth has three main layers: the crust, mantle and the inner and outer core
- [ ] the Earth's atmosphere has changed over millions of years. Many of the gases that make up the atmosphere came from volcanoes
- [ ] for 200 million years, the proportions of different gases in the atmosphere have been much the same as they are today
- [ ] human activities have recently produced further changes, e.g. the levels of greenhouse gases are rising

P1a Energy and electricity

I know:

how heat (thermal energy) is transferred and what factors affect the rate at which heat is transferred

- [ ] heat energy can be transferred by conduction, convection and thermal radiation
- [ ] thermal conductors (e.g. metals) transfer heat energy easily; thermal insulators (e.g. plastic, glass) do not
- [ ] dark, dull surfaces emit and absorb thermal radiation better than shiny, light surfaces
- [ ] the bigger the temperature difference between an object and its surroundings, the faster the rate at which heat is transferred

what is meant by the efficient use of energy

- [ ] energy is never created nor destroyed; some energy is usually wasted as heat
- [ ] Sankey diagrams show energy input and output of a device

why electrical devices are so useful

- [ ] they transform electrical energy to whatever form of energy we need at the flick of a switch
- [ ] the National Grid transmits energy around the country at high voltages and low current to reduce energy losses
- [ ] dynamos produce electricity when coils of wire rotate inside a magnetic field
- [ ] how to work out the power rating of an appliance (the rate at which it transforms electrical energy)
- [ ] to calculate the amount of energy transferred from the mains use:
  energy transferred = power × time
  (kilowatt-hour, kWh) (kilowatt, kW) (hour, h)
- [ ] to calculate the cost of energy transferred from the mains use:
  total cost = number of kilowatt-hours × cost per kilowatt-hour

how we should generate the electricity we need

- [ ] we need to use more renewable energy sources, including wind, hydroelectric, tidal, wave and geothermal power
- [ ] most types of electricity generation have some harmful effects on people or the environment; there are also limitations on where they can be used

P1b Radiation and the Universe

I know:

what the uses and hazards of the waves that form the electromagnetic spectrum are

- [ ] from longest to shortest wavelength: radio waves, microwaves, infrared, visible light, ultraviolet, X-rays, gamma rays
- [ ] radiation has many uses e.g. radio, TV, satellites, cable and mobile phone networks
- [ ] some forms of electromagnetic radiation can damage living cells: ionising radiation (ultraviolet, X-rays and gamma rays) can cause cancer

what the uses and dangers of emissions from radioactive substances are

- [ ] the uses and hazards of radioactive substances (which emit alpha particles, beta particles and gamma rays) depend on the wavelength and frequency of the radiation they emit
- [ ] background radiation is all around us e.g. granite rocks can emit gamma rays and form radioactive radon gas
- [ ] the relative ionising power, penetration through materials and range in air of alpha, beta and gamma radiations

about the origins of the Universe and how it continues to change

- [ ] the Universe is still expanding; in the beginning, matter and space expanded violently and rapidly from a very small initial point i.e. the Big Bang
- [ ] red shift indicates that galaxies are moving apart; the further away a galaxy, the faster it is moving away from us
- [ ] telescopes on Earth and in space give us information about the Solar System and the galaxies in the Universe

Checklists – Additional Science Unit 2

B2a Discover Martian living!

I know:

what animals and plants are built from

☐ animal cells and plant cells have a membrane, cytoplasm and a nucleus; plant cells may also have a vacuole, chloroplasts and a cell wall

☐ in multicellular organisms, different cells are specialised for different functions

☐ the chemical reactions inside cells are controlled by enzymes

how dissolved substances get into and out of cells

☐ diffusion is the net movement of particles of gas, or substances dissolved in a solution, from a region of high concentration to a region of lower concentration

☐ osmosis is the diffusion of water molecules through a partially permeable membrane

how plants obtain the food they need to live and grow

☐ green plants use chlorophyll to trap energy from the Sun to photosynthesise

☐ leaves are specially adapted for photosynthesis – they can be broad, flat, thin and have lots of stomata

☐ the rate of photosynthesis is affected by light intensity, carbon dioxide concentration and temperature

☐ mineral salts in the soil are used to make proteins or chlorophyll

what happens to energy and biomass at each stage in a food chain

☐ energy passes along food chains but some energy is lost at every stage

☐ the shorter the food chain, the less energy is lost

☐ the mass of biomass at each stage in a food chain is less than it was at the previous stage. This can be shown in a pyramid of biomass

☐ reducing energy loss increases efficiency of food production

☐ decomposers and detritus feeders feed on dead organisms and their waste

B2b Discover DNA!

I know:

what enzymes are and what their functions are

☐ enzymes are proteins that act as biological catalysts, speeding up chemical reactions

☐ each enzyme works at an optimum temperature and pH

☐ they are involved in respiration, photosynthesis, protein synthesis and digestion

☐ enzymes are used in washing powders and in industry

how our bodies keep internal conditions constant

☐ blood sugar levels are controlled by the pancreas, which makes insulin to bring down blood sugar levels

☐ sweating cools the body down and helps to maintain a steady body temperature

☐ waste products e.g. carbon dioxide and urea must be removed from the body

some human characteristics show a simple pattern of inheritance

☐ some inherited characteristics are controlled by a single gene

☐ different forms of a gene are called alleles; in homozygous individuals the alleles are the same, in heterozygous individuals they are different

☐ in mitosis each new cell has the same number of identical chromosomes as the original

☐ sex chromosomes determine the sex of the offspring (male XY, female XX)

☐ stem cells can specialise into many types of cells

C2a Discover Buckminsterfullerene!

I know:

how sub-atomic particles help us to understand the structure of substances

☐ atoms consist of sub-atomic particles called protons, neutrons and electrons

☐ an element's mass number is the number of protons plus the number of neutrons in an atom

☐ an element's atomic number is the number of protons in an atom

☐ the arrangement of electrons can be used to explain what happens when elements react and how atoms join together to form different types of substances

how structures influence the properties and uses of substances

☐ ionic bonding is the attraction between oppositely charged ions

☐ ionic compounds are giant lattice structures with high melting points that conduct electricity when molten or dissolved

☐ non-metal atoms can share pairs of electrons to form covalent bonds

☐ giant covalent structures are macromolecules that are hard, have high melting points but do not conduct electricity

☐ nanoparticles are very small structures with special properties because of their unique atom arrangement

how much can we make and how much we need to use

☐ the relative masses of atoms can be used to calculate how much to react and how much we can produce, because no atoms are gained or lost in chemical reactions

☐ the percentage of an element in a compound can be calculated from the relative masses of the element in the formula and the relative formula mass of the compound

☐ high atom economy (atom utilisation) is important for sustainable development and economic reasons

☐ in some chemical reactions, the products of the reaction can react to produce the original reactants; they are called reversible reactions

C2b Discover electrolysis!

I know:

how we can control the rates of chemical reactions

☐ the rate of a chemical reaction can be found by measuring the amount of a reactant used or the amount of product formed over time

☐ chemical reactions can be speeded up by increasing the: temperature; concentration of a solution; pressure of a gas; surface area of a solid; and by using a catalyst

☐ particles must collide with sufficient energy in order to react; the minimum energy required is the activation energy

whether chemical reactions always release energy

☐ chemical reactions involve energy transfers

☐ exothermic reactions give OUT energy; endothermic reactions take IN energy

☐ if a reversible reaction is exothermic in one direction, it is endothermic in the opposite direction. The same amount of energy is transferred in each case

how can we use ions in solutions

☐ when molten or dissolved in water, ions in ionic compounds are free to move around

☐ passing an electric current through an ionic compound breaks it down into its elements: this is called electrolysis

☐ at the negative electrode, positively charged ions gain electrons (reduction) and at the positive electrode, negatively charged ions lose electrons (oxidation)

☐ electrolysis of sodium chloride solution makes hydrogen, chlorine and sodium hydroxide; copper can be purified by electrolysis

☐ metal oxides and hydroxides are bases and react with acids to form salts

☐ soluble salts can be made from acids and insoluble salts can be made by mixing solutions of ions

☐ in neutralisation reactions, hydrogen ions from acids react with hydroxide ions to produce water

P2a Discover forces!

I know:

how we can describe the way things move

☐ distance-time graphs show how an object moves

☐ velocity-time graphs show how the velocity changes; the slope represents acceleration, and the area under the graph represents distance travelled

how we make things speed up or slow down

☐ an unbalanced force acting on an object changes its speed

☐ forces can add up or cancel out to give a resultant force; when the resultant force is *not* zero, an object accelerates:
resultant force (N) = mass (kg) × acceleration (m/s^2): $F = ma$

☐ an object falling through a fluid accelerates until it reaches a terminal velocity, when the resultant force *is* zero

☐ the stopping distance of a car is the thinking distance plus the braking distance. This increases as the speed increases

what happens to the movement energy when things speed up or slow down

☐ when a force causes an object to move, energy is transferred and work is done

☐ every moving object has kinetic energy that can be transformed into other forms

☐ the kinetic energy of a body depends on its mass and its speed

what momentum is

☐ every moving object has momentum that depends on its mass and its velocity:
momentum (kg m/s) = mass (kg) × velocity (m/s)

☐ momentum has size and direction and is conserved in collisions and explosions

what static electricity is, how it can be used and the connection between static electricity and electric currents

☐ rubbing electrical insulators together builds up static electricity because electrons are transferred

☐ if an object gains electrons it has a negative charge; if it loses electrons it has a positive charge

☐ electrostatic charges can be used in photocopiers, smoke precipitators and paint sprayers

☐ when electrical charges move we get an electrical current

P2b Discover nuclear fusion!

I know:

what the current through an electrical current depends on

☐ the symbols for components shown in circuit diagrams

☐ resistance is increased in long, thin, heated wires

☐ the current through a component depends on its resistance. The greater the resistance the smaller the current for a given p.d. across the component

☐ **potential difference = current × resistance**
(volt, V) (ampere, A) (ohm, Ω)

☐ in a series circuit: the total resistance is the sum of the resistance of each component; the current is the same through each component and the total p.d. of the supply is shared between the components

☐ in a parallel circuit: the total current through the whole circuit is the sum of the currents through the separate components; the p.d. across each component is the same

what mains electricity is and how it can be used safely

☐ mains electricity is an a.c. supply of 230 V and has a frequency of 50 Hz; it is very dangerous

☐ fuses and earth wires protect appliances from damage and people from harm or death

☐ three-pin plugs must be wired correctly and hold the correct fuse

why we need to know the power of electrical appliances

☐ the power of an electrical appliance is the rate at which it transforms energy: $$\text{power (W)} = \frac{\text{energy transformed (J)}}{\text{time taken (s)}}$$

☐ power, potential difference and current are related by the equation:
power = current × potential difference
(watt, W) (ampere, A) (volt, V)

☐ most appliances have their power and the p.d. they need on them so we can calculate the current and fuse required

what happens to radioactive substances when they decay

☐ isotopes (elements with the same number of protons but a different number of neutrons) with unstable nuclei emit energy as particles or rays called radiation

☐ the Rutherford and Marsden scattering experiment revealed the structure of the atom

☐ background radiation comes from rocks, soil, cosmic rays, living things and medical X-rays

what nuclear fission and nuclear fusion are

☐ nuclear fission is the splitting of an atomic nucleus; it is used in nuclear reactors

☐ nuclear fusion is the joining of two smaller nuclei to form a larger one; stars release energy by nuclear fusion

Answers

ANSWERS

Unit B1a Human biology
Page 4
1 Brain and spinal cord.
2 Motor neurone.
3 A flash of light; a smell; a loud noise (you may think of others).
4 Adrenaline: heart, breathing muscles, eyes and digestive system. Reproductive hormones/oestrogen: ovaries.

Page 5
1 Sensory, relay and motor.
2 The neurone secretes a chemical, which diffuses across the gap and produces an impulse in the next neurone.
3 To provide cells with a constant supply of energy.
4 The water evaporates and takes heat from the skin.

Page 6
1 Ovary.
2 It makes eggs mature in the ovaries; it makes the ovaries secrete oestrogen.
3 They contain oestrogen, which stops FSH being produced, so no eggs develop.
4 Advantages: fewer unwanted pregnancies; fewer abortions; slower population growth. Disadvantages: could encourage more sexual partners; could increase spread of sexually transmitted diseases.

Page 7
1 Carbohydrates, fats, proteins, vitamins, minerals, roughage, water.
2 The greater the muscle-to-fat ratio, the faster the metabolic rate.
3 Very overweight.
4 Bones rub together as the joint moves, which is very painful. The joint may become too stiff to move.

Page 8
1 Drought, floods, war.
2 Lack of protein in the diet.
3 Unsaturated fats.
4 HDLs.

Page 9
1 All of them.
2 They experience withdrawal symptoms.
3 It is tested on tissues or animals in a laboratory.
4 Neither the experimenters nor the subjects know who is getting the drug and who is getting a placebo.

Page 10
1 Heroin, cocaine.
2 If they inject the drug and share needles, they may introduce the AIDS virus into their body.
3 It slows down reactions and may cause brain cells to shrink.
4 The liver is often the first to be really badly damaged. The brain may also be harmed.

Page 11
1 Nicotine.
2 There seems to be a link between them. When one increases, so does the other. It does not, though, necessarily mean that one *causes* the other.
3 A microorganism (e.g. a bacterium or a virus) that causes disease.
4 Doctors were no longer carrying pathogens from one patient to another.

Page 12
1 A chemical produced by white blood cells that sticks to antigens (e.g. bacteria) and destroys them. Antibodies are specific to a particular kind of antigen.
2 White blood cells, in particular lymphocytes.
3 They kill bacteria in the body without harming our cells.
4 Viruses reproduce inside body cells, so it is difficult to destroy the viruses without destroying the cells as well.

Page 13
1 A bacterium. (It is resistant to many antibiotics, including methicillin.)

2 It was a new disease, so no-one had immunity to it and there were no drugs to combat it. International travel meant it could spread quickly all over the world.
3 A small quantity of dead or weakened pathogens is injected into the blood stream. Your white blood cells make antibodies against it, so if you are infected with it in the future your immune system will destroy it immediately.
4 Many parents did not let their children have the jab, so they were not immune against measles, mumps or rubella. They could have got one of these diseases or helped to spread it to others.

Unit B1b Evolution and environment
Page 15
1 Cold climate.
2 Arctic; heat exchange takes place between warm blood in arteries and cold blood in veins in the legs.
3 Food, territory, mates.
4 Warns predators not to eat them.

Page 16
1 They are genetically identical.
2 Sexual involves gametes, asexual does not. Sexual involves fertilisation, asexual does not. Sexual produces variation in the offspring, asexual does not. (NB – it is not true to say that sexual reproduction always involves two parents, as there are many plants where one flower produces both male and female gametes and they fertilise themselves.)
3 On chromosomes in the nucleus of a cell.
4 By genes inherited from your parents.

Page 17
1 A piece cut from a plant from which a whole new plant can grow.
2 The new plants are all genetically identical to the parent. They all look the same. They all grow at the same rate. It is a quick and cheap way of getting many new plants.
3 Tissue culture.
4 A nucleus from the cell to be cloned is placed inside an egg cell from which the nucleus has been removed.

Page 18
1 It has been given genes from a different kind of organism.
2 Bacterium.   3 Lamarck and Darwin.
4 They have become adapted for eating different kinds of food.

Page 19
1 Mutation.
2 The black individuals were less likely to be eaten by birds, because they were better camouflaged. So they were more likely to reproduce, passing on the gene for black coloration to their offspring.
3 A dead organism is buried and compressed in sediment.
4 They have adaptations that suggest they lived in swampy ground.

Page 20
1 Its eggs were eaten by animals introduced to Mauritius by humans.
2 The species that is the best competitor may prevent the original species from getting enough resources to survive.
3 We produce more waste as our population increases and living standards improve.
4 The rate of increase is getting less (the graph is getting flatter).

Page 21
1 Quarrying for building materials, building houses on green sites, farming, dumping waste.
2 Getting as much production from crops or animals as possible in a small area, using lots of fertilisers and pesticides.
3 Untreated sewage, fertilisers, toxic chemicals.
4 It does not break down. It became concentrated up the food chain, so predators ended up with such large amounts in their bodies that it killed them.

Page 22
1 Sulfur dioxide.
2 It can trigger asthma attacks or cause bronchitis.
3 Sulfuric acid, nitric acid, carbonic acid.
4 It reacts with calcium carbonate, causing the stone to break down.

Page 23
1 Lichens.   2 Heavily polluted.
3 There is less photosynthesis, so less carbon dioxide is taken from the air and stored in wood. So carbon dioxide levels in the atmosphere can increase.
4 The microorganisms that are breaking them down respire, releasing extra carbon dioxide into the air.

Page 24
1 Carbon dioxide, methane.
2 They have both increased.
3 Improving the quality of life without compromising future generations.
4 Nuclear power; wind power; wave power; hydroelectric power.

Unit C1a Products from rocks
Page 26
1 One.   2 Zn
3 Neutrons and protons.
4 12. It has 12 protons in its nucleus, and 12 electrons orbiting it.

Page 27
1 An element.   2 Two more.
3 Calcium carbonate.
4 Advantages: jobs; roads; local income. Disadvantages: habitat destruction; noise; traffic (there may be other good suggestions).

Page 28
1 calcium carbonate → calcium oxide (quicklime) + carbon dioxide
2 $CaCO_3 \rightarrow CaO + CO_2$
3 By heating sand, limestone and sodium carbonate together.
4 Concrete is a mixture of cement, sand and rock chippings; it is stronger than cement.

Page 29
1 Iron oxide.
2 Oxygen is removed from it; the oxygen is taken by carbon, because carbon is more reactive than iron.
3 Impurities are removed from it.
4 They are made of a lattice of atoms all of the same size, which form layers and can easily slide over each other.

Page 30
1 Always carbon, and sometimes other metals.
2 It is very hard (and can be sharpened to a fine edge).
3 Only cobalt is a transition metal.
4 It has delocalised electrons which can move freely.

Page 31
1 Bauxite.
2 Aluminium is more reactive than carbon, so carbon is not able to take oxygen away from the aluminium.
3 Saves energy used in extracting aluminium from bauxite; reduces need for bauxite mines; reduces need to dispose of waste in landfill sites (there are other possible suggestions).
4 Destroys habitats; pollutes the air with carbon dioxide; may spread diseases between migrant workers (there are other possible suggestions).

Page 32
1 It is light (has a low density), strong and resists corrosion.
2 Titanium dioxide does not conduct electricity (because it is a covalent compound).
3 It is a good conductor of electricity and can easily be drawn out into wires.
4 There is only a limited supply of high-grade ores, so if we go on using copper we will need to get it from low-grade ores.

Page 33
1 Gold is very unreactive.
2 They can go back to their original shape after being deformed. Example: copper-zinc-aluminium alloy; nickel-titanium alloy.
3 The supplies are finite. We are using it up faster than any new oil is formed.
4 It does not produce carbon dioxide or sulfur dioxide when it burns.

Page 34
1 Hydrogen and carbon.
2 Those with small molecules.
3 C_2H_6.   4 C_3H_8.

Page 35
1 More sulfur dioxide will be formed, which causes acid rain.
2 It contains carbon dioxide, which is found in the air and dissolved in rain drops to form carbonic acid.
3 It prevents acid rain from being produced; sulfur is valuable and can be made use of.
4 The gases are sprayed with a slurry of limestone, which reacts with the sulfur dioxide to form solid calcium sulfate.

Unit C1b Oils, Earth and atmosphere
Page 37
1 They do not contain any C–C double bonds.
2 They are heated to form a vapour and passed over a catalyst.
3 They contain at least one C–C double bond.
4 The double bonds can break and become single bonds, leaving a 'spare' bond to react with something else.

Page 38
1 A mixture of ethene and steam is passed over a catalyst.
2 It has an –OH group.
3 A substance made of many smaller molecules, called monomers, joined together in a long chain.
4 They have C–C double bonds, which can 'open' and react with another alkene molecule.

Page 39
1 Slippery; non-toxic; inert (unreactive); withstands high temperatures.
2 Viscosity.
3 They are not biodegradable.
4 Mix them with cellulose or starch, which microorganisms can break down.

Page 40
1 Fruits; nuts; seeds. (From named plants.)
2 Dissolve the crushed plant material in a hydrocarbon solvent. Distillation may also be used.
3 Rapeseed oil; sunflower oil; wood (there are others, e.g. elephant grass).
4 Non-renewable fossil fuels are running out.

Page 41
1 Mayonnaise; milk; emulsion paint (there are others).
2 Light is reflected from the tiny droplets of the dispersed phase.
3 A saturated fat contains no C–C double bonds, but an unsaturated fat does.
4 Add it to bromine water. If the bromine water changes from orange to colourless, the liquid contains unsaturated molecules.

Page 42
1 Vegetable oils.
2 Extra hydrogen atoms are added to the C–C double bonds, so the molecules become straight instead of bent. They can pack together more tightly, forming a solid.
3 A substance added to food to improve its taste, appearance or keeping qualities.
4 The E-number identifies the additive, which will have been tested to make sure it is safe.

Page 43
1 Chromatography.
2 Spots of the colour are placed on a pencil line on paper, which stands in a solvent. The solvent moves up the paper, taking the colours with it. Some colours travel faster than others. The height of the spot of an unknown colour can be matched with heights of spots of reference colours.

3 Probably about 4600 million years (4.6 billion years).
4 Crust, mantle and core.

Page 44
1 They thought that the surface of the Earth cracked as it cooled.
2 Heat generated by radioactive processes deep inside the Earth. (There is also heat left over from when the Earth was first formed.)
3 Tension builds up where two plates are moving in relation to each other. Then suddenly the two plates slip.
4 At a subduction zone, an oceanic plate is gradually moving underneath a continental plate. Friction stops the plates moving smoothly. The plates move in sudden jerks, causing earthquakes.

Page 45
1 Nitrogen, oxygen and water vapour.
2 0.04% (or 0.035%).
3 Argon, neon, helium, krypton and xenon.
4 Living things produced oxygen when they photosynthesised.

Page 46
1 It has increased by 1 °C.
2 It is a greenhouse gas, preventing heat from being lost into space.
3 A place where carbon is locked up. Sedimentary rocks (especially limestone), plants (especially trees) and fossil fuels.
4 Carbon dioxide dissolves in sea water to form carbonic acid, which reacts with calcium to form insoluble calcium carbonate. This collects on the sea floor, and gradually becomes compressed to form solid rocks.

Unit P1a Energy and electricity
Page 48
1 From the coffee (hotter) to your hands (colder).
2 The person is hotter than the trees, and so emits more thermal radiation which is detected by the camera.
3 More.
4 Silver surfaces emit and absorb thermal radiation more slowly than black surfaces.

Page 49
1 It is less dense than cold air.
2 Hotter particles vibrate faster. They are in close contact with their neighbours, so they make them vibrate faster as well.
3 Air in between the layers is a poor conductor. So less heat is conducted away from your warm body to the cold air outside your clothes.
4 Their particles are very far apart and only rarely come into contact with each other.

Page 50
1 Potential (gravitational potential).
2 10 million J.
3 Energy transfer: kinetic energy in arm to kinetic energy in gong.
Energy transformation: kinetic energy in gong to sound energy in air.
4 Heat energy in the bike, the surrounding air and the ground.

Page 51
1 400 J.
2 Exhaust gases: 215 J; moving parts: 510 J.
3 Conduction, convection and radiation.
4 The heat energy in the bath water is spread out through a much larger volume than the heat energy in the kettle water.

Page 52
1 Each bag weighs $2 \times 10 = 20$ N. So work done = $20 \times 20 \times 3 = 1200$ J
2 $1200 \div 60 = 20$ W   3 Machine A.
4 $(500 \div 1000) \times 100 = 50\%$

Page 53
1 It reduces the quantity of hot air leaving the room, so less heat energy needs to be used to keep the air in the room warm.
2 The fluorescent bulb, because it only loses 80% of its energy as heat.
3 Chemical (not electrical – you cannot store electrical energy.)
4 Usually because there is no mains electricity supply available. (It may also be cheaper.)

Page 54
1 The fuse gets hot and melts.
2 A thin wire has more resistance than a thick one. The more resistance, the more it heats up.
3 50 J.
4 number of kWh = power (kW) $\times$ time (hours)
number of kWh = $(50 \div 1000) \times 0.5 = 0.025$ kWh
cost = 8p $\times$ 0.025 = 0.2p

Page 55
1 A step-down transformer.
2 High voltage means low current, and this means less energy is lost as heat and the wires can be thinner.
3 An electric motor uses electricity to produce movement. A dynamo uses movement to produce electricity.
4 If either the magnet or coil of wire are moved relative to each other, a current will be produced in the wire.

Page 56
1 A turbine comes before a generator in a power station. A turbine is made to turn, using an energy source such as steam. The turbine then turns the generator, which generates electricity.
2 35%.
3 The water that is used is renewed when it rains, and as rivers flow into the reservoir.
4 The steam can turn a turbine connected to a generator; this is geothermal energy.

Page 57
1 They may think they look ugly; some people think they are noisy; birds may fly into them and be killed, which may upset people.
2 So that the fuel can easily be brought to them by road in large vehicles.
3 They are running out; they generate carbon dioxide which may be causing global warming.
4 They rely on water flowing rapidly downhill; they need to be in areas where there is high rainfall.

Unit P1b Radiation and the Universe
Page 59
1 Energy.
2 Radar (speed detectors); telecommunications (satellites, mobile phones).
3 Radio waves.
4 Short wavelength.

Page 60
1 (Visible) light.
2 Gamma rays, X-rays, ultraviolet, visible light, infrared, microwaves, radio.
3 It absorbs the energy in the microwaves.
4 Most go straight through, but some are absorbed.

Page 61
1 Ultraviolet.
2 It can ionise atoms and molecules, damaging DNA; this can cause cancer.
3 A signal that is sent as pulses (a series of 'ons' and 'offs').
4 A satellite that always stays above the same spot on the Earth's surface.

Page 62
1 Glass.
2 They undergo total internal reflection when they hit the edge of the fibre.
3 Yes, because they both contain the same number of protons.
4 The radiation comes from the nucleus of an atom.

Page 63
1 Alpha, beta and gamma.
2 Alpha radiation cannot penetrate skin, but it has a very high ionising effect which could damage atoms inside you if swallowed.
3 Radon gas; rocks and soil.
4 The badge shows how much radiation they have been exposed to, so they can check that they stay within safe limits.

Page 64
1 12 years.
2 The count rate will halve.
3 A radioisotope is added to the water in the pipe, and a Geiger counter can detect where most radiation is present.
4 Alpha would not go through even the thinnest paper. Gamma would go through even the thickest paper.

Page 65
1 It ionises atoms, and may damage DNA; a cell in which this has happened may divide uncontrollably.
2 The waste is radioactive, so it could harm people, e.g. by damaging their DNA.
3 The light that reaches it does not have to pass through the atmosphere, which can cause distortion.
4 More light can enter a reflecting telescope.

Page 66
1 A pull force between objects.
2 They used the gravity of Jupiter and Saturn to speed them up as they flew past.
3 The joining together of the nuclei of two atoms.
4 They become helium atoms and release energy.

Page 67
1 Jupiter, Saturn, Uranus and Neptune.
2 The gases were pushed further away than rocks when the early star exploded.
3 It will become a black dwarf.
4 There is no nuclear fusion, so no radiation pressure. So there is no outward force to act against gravity, which pulls the particles together.

Page 68
1 A huge amount of energy at a single point exploded, forming protons, neutrons and electrons.
2 Hydrogen.
3 If something is moving away from us, the frequency of the light coming to us appears to be less.
4 It shows that the galaxies are moving away from each other, and that the further away they are the faster they are moving; this could be explained if everything was spreading outwards from a single point after the Big Bang.

Unit B2a Discover Martian living!
Page 70
1 Chloroplast; cell wall; permanent vacuole.
2 Ribosome.
3 Tail for swimming; lots of mitochondria for energy; nucleus with only one set of chromosomes; enzymes in head to digest into egg.
4 In the linings of the alimentary canal and gas exchange system; they make mucus.

Page 71
1 Solution, gas.
2 It decreases the rate, because particles move more slowly.
3 Oxygen.
4 The greater the concentration gradient, the faster the rate of diffusion.

Page 72
1 Water.   2 Dilute solution.
3 Cell membrane.   4 Cell wall.

Page 73
1 carbon dioxide + water (+ light energy)
   → glucose + oxygen
2 Respiration.
3 Carbon dioxide can diffuse through it more quickly. Light can reach all the cells in the leaf.
4 They are near the top of the leaf, where they get sunlight. They are next to the spongy mesophyll layer, where there is carbon dioxide in the air spaces.

Page 74
1 Higher light intensity, higher carbon dioxide concentration, higher temperature (so long as it is not too high).

2 The plant is using as much light as it can; it may be limited by the amount of chlorophyll it contains, or by not having enough carbon dioxide to photosynthesise any faster.
3 Chlorophyll.
4 The nitrogen in the ions is combined with glucose to make amino acids, which are strung together to make proteins.

Page 75
1 Energy flow.
2 Some of the sunlight does not fall onto their leaves, some is reflected from the leaves, some passes straight through the leaves.
3 The mass of living organisms.
4 Mammals use a lot of food to produce heat energy to keep their body temperature constant. Reptiles just take on the heat of the surroundings.

Page 76
1 The food chain is shorter, so less energy is lost.
2 The animals need less food to produce heat energy to keep themselves warm.
3 They use central distribution points and large container lorries.
4 The farmer gets fewer eggs per hen or per square metre of land.

Page 77
1 They produce enzymes that break down waste plant and animal material.
2 Reactions happen faster at higher temperatures.
3 Earthworm.
4 Every organism, when it dies.

Page 78
1 Carbohydrates; fats; proteins; chlorophyll.
2 To reduce carbon dioxide levels in the atmosphere; to produce oxygen for us to breathe; to provide food.
3 By respiration.
4 They absorb it during photosynthesis, and store carbon in their growing bodies as carbon compounds.

Unit B2b Discover DNA!
Page 80
1 A place in an enzyme molecule where its substrate fits.
2 The enzyme molecules are denatured – they lose their shape, so the substrate does not fit in the active site.
3 It digests starch molecules to sugar molecules (maltose).
4 It is an alkali, and neutralises the acidic contents arriving from the stomach. It contains bile salts, which emulsify fats so that lipase can digest them more easily.

Page 81
1 They digest the protein (haemoglobin) in blood and make it soluble so it can be washed away.
2 They break up the stain and lift it from the cloth, so the enzymes can easily get at the fat molecules.
3 It tastes sweeter than other sugars, so less needs to be added.
4 They break down proteins into amino acids, which babies can absorb.

Page 82
1 glucose + oxygen → carbon dioxide + water
2 They keep their body temperature constant, usually above the temperature of their environment; this requires heat energy, which they release from food by respiration.
3 Intercostal muscles and diaphragm.
4 Large surface area, thin walls, rich blood supply.

Page 83
1 Liver.
2 A tiny tubule in a kidney.
3 Temperature; glucose concentration; water content; ion concentration.
4 To conserve heat; because less water may have been lost in sweat.

Page 84
1 In the brain.
2 Their thermoregulatory centre stops working when it is very cold.
3 Pancreas.

4 Banting and Best injected fluid containing it into dogs with diabetes, which became healthy.

Page 85
1 Two.   2 Mitosis.
3 So that when they fuse together the new cell will have the normal two sets.
4 They contain mixtures of alleles (different varieties of a gene) from both their parents.

Page 86
1 A cell that is not differentiated, and that can form different types of specialised cells.
2 White blood cells; red blood cells; bone cells; cartilage cells; tendon cells; fat cells.
3 DNA.
4 So that each daughter cell receives a complete set of chromosomes/DNA.

Page 87
1 Always producing offspring like themselves.
2 An allele that does not show unless the dominant allele is absent.
3 The sperm.
4 There equal numbers of sperm with a X chromosome and sperm with a Y chromosome, so there is an equal chance that the zygote will get an X or a Y from the sperm.

Page 88
1 Cystic fibrosis and Huntington's disease.
2 If they are both heterozygous, they could each pass on the recessive gene to a child.
3 Everyone's DNA is unique (except for identical twins).
4 Linking suspects to the scene of a crime; checking who is the father of a child.

Unit C2a Discover Buckminsterfullerene!
Page 90
1 Proton +1, neutron zero, electron –1.
2 39. It has a total of 39 neutrons and protons in its nucleus. (You can also work out how many of each it has. Its atomic number is 19, so it has 19 electrons and 19 protons. So it must have 20 neutrons.)
3 One.
4 By losing this one electron.

Page 91
1 Three. The atom has no overall charge.
2 They all have the same number of electrons in their outer shell, which gives them similar chemical properties.
3 The potassium atom can become stable by losing one electron, and the chlorine atom can become stable by gaining one.
4 The total charge on the positive ions is exactly balanced by the total charge on the negative ions.

Page 92
1 Molten (liquid) or in solution.
2 A compound made up of many ions all arranged in a regular lattice, held together by strong forces of attraction between the oppositely charged ions.
3 Covalent.
4 They each have a complete outer electron shell.

Page 93
1 Hydrogen; nitrogen; oxygen; chlorine.
2 They do not contain any charged particles.
3 Carbon.
4 The carbon atoms are arranged in layers so the layers can slide over one another.

Page 94
1 Their atoms lose electrons, forming positive ions.
2 Their layers of atoms can easily slide over each other.
3 One.
4 Rubidium, because metals become more reactive as you go *down* the Group.

Page 95
1 The elements in Group 7.
2 They gain one electron to form negative ions.
3 They are tiny particles made up of just a few hundred atoms.
4 They could be used as new types of catalysts.

Page 96
1 It is able to revert to its original state, after being changed by a change in its environment.
2 It darkens; it will then go back to its normal colour when the light is no longer present.
3 There is one carbon atom and four hydrogen atoms, chemically bonded together.
4 In a compound, the elements are chemically combined and they are in fixed proportions. In a mixture, they are not chemically combined and they can be in any proportion.

Page 97
1 $12 + (4 \times 1) = 16$ 2 $\dfrac{12}{16} \times 100 = 75\%$
3 16 g 4 One.

Page 98
1 They are the same. No atoms are lost or created.
2 The reaction may be reversible. Some of the product may be lost when it is separated from other products. Some of the reactants may react in unexpected ways.
3 $\rightleftharpoons$
4 nitrogen + hydrogen $\rightleftharpoons$ ammonia

Page 99
1 The forward reaction and backwards reaction happen at the same time – so there are always some reactants present.
2 It has no effect (but it does speed up the rate of the reaction).
3 Ammonia. 4 Iron.

Unit C2b Discover electrolysis!
Page 101
1 Use less concentrated acid, use cooler acid, use a thicker piece of magnesium (with less surface area).
2 Measuring the volume of a gas produced during a reaction.
3 The mass decreased as the hydrogen produced was lost to the air.
4 Time.

Page 102
1 There are more particles in the same space, so they collide more often.
2 It would double it (because particles would collide twice as frequently).
3 Sulfur is formed, which makes the mixture cloudy.
4 It is twice as fast. (At 20 °C it took 500 s for the cross to disappear, and at 30 °C it took 250 s.)

Page 103
1 Tiny chips.
2 By that time, the reaction would be complete, and there is no difference in the *total* mass of gas lost at the end of the reaction.
3 Hydrogen.
4 There are twice as many acid particles in the same volume, so the collisions between the acid particles and the magnesium particles will be twice as frequent.

Page 104
1 Manganese(IV) oxide (*or* the enzyme catalase).
2 Transition metals.
3 Exothermic (heat is given out).
4 The amount of energy taken in when the reaction goes one way is the same as the amount of energy given out when the reaction goes the other way.

Page 105
1 Equilibrium.
2 Cool things down.
3 Making nitrate fertilisers.
4 High pressure (200 atmospheres), moderate temperature (about 450 °C), iron catalyst.

Page 106
1 As a liquid (molten) or in solution.
2 Lead ions are positively charged, and opposite charges attract.
3 'Sodium' is not positive – it should say 'Sodium ions'.
4 They lose their electron and become chlorine atoms. This is oxidation – the loss of electrons.

Page 107
1 Hydrogen.
2 As a fuel; making ammonia (for fertilisers); making margarine; in weather balloons.
3 Hydrochloric acid.
4 To get rid of any unreacted pieces of zinc.

Page 108
1 Magnesium chloride and water.
2 Zinc oxide and sulfuric acid.
3 How acidic or alkaline it is – its pH.
4 Using the indicator tells you exactly how much base and acid to react together. You can then use these quantities in another reaction, to get a sample of the salt that is not contaminated with indicator.

Page 109
1 H^+
2 $H^+(aq) + OH^-(aq) \rightarrow H_2O(l)$
3 A solid that drops to the bottom of a liquid; a precipitate forms when ions in solution react to form an insoluble compound.
4 The calcium ions react with the phosphate ions to form insoluble calcium phosphate, which precipitates out.

Unit P2a Discover forces!
Page 111
1 C; the line is steepest. 2 14 km/h
3 Speed is the distance travelled in a certain time. Velocity is the same, but it includes the direction of travel as well.
4 1 m/s^2

Page 112
1 The line is horizontal (*not* 'straight').
2 The line starts off part way up the *x*-axis, then horizontal, then sloping downwards right to the bottom of the graph, then sloping upwards, then horizontal again at the same level at which it began.
3 The resultant force is 0, so the parachute continues to move at a constant velocity.
4 The box does not move, because the resultant force is 0.

Page 113
1 The small mass. 2 … directly …
3 … opposite … 4 … accelerate.

Page 114
1 The fastest, steady speed at which it falls.
2 It increases air resistance, so the resultant downward force is smaller.
3 … thinking …; … braking …
4 Having a slower reaction time (e.g. because of drinking alcohol; taking drugs; being tired); driving faster.

Page 115
1 Air resistance is the frictional force caused by movement in air; drag is the same, but in water.
2 The resistance against movement in water (drag) is greater than in air (air resistance) because water is denser than air.
3 It is transformed to sound energy, and to heat energy in the ground (and the air).
4 a At the top of the swing; b at the bottom.

Page 116
1 Joules.
2 30 J (15 × 1 for each parcel).
3 The lorry, because its mass is larger.
4 The elastic potential energy stored in the elastic band.

Page 117
1 … velocity.
2 Momentum has a direction (it is a vector quantity) whereas kinetic energy does not (it is a scalar quantity).
3 Zero (because velocity is zero).
4 The momentum of the balloon and the momentum of the air add up to zero (they are moving in different directions).

Page 118
1 Yes, because the direction has changed.
2 The seat belt slows down the rate of change of momentum when you suddenly slow down to a stop. The longer the time taken, the less force is used. Less force means less injury.
3 Static (electricity).
4 Electrons have moved from the shirt to the tent fabric.

Page 119
1 The charge can flow away through the conductor.
2 The strip gives electrons to the cap, and these move down the rod and the leaf. Now the rod and leaf have the same negative charge, so they repel each other.
3 They all have a positive charge, so they repel each other.
4 The charge flows away to earth – the charged object is discharged.

Page 120
1 A spark jumping from one plane to the other could cause the fuel to explode.
2 Any charge that builds up can be conducted safely away along the metal hose.
3 Opposite charges on the particles of paint and the car attract the paint to the car's surface, rather than spreading out into the air.
4 The drum had positive charge, but it only keeps its charge in the places corresponding to the dark parts of the object being photocopied. (The light parts make the drum able to conduct, so the charge flows away.)

Unit P2b Discover nuclear fusion!
Page 122
1 Check your answer against the symbols on page 122.
2 Your circuit should look just like the first circuit diagram on page 122, except that it contains another ammeter somewhere in the circuit, and only one lamp with the voltmeter going across it.
3 Electrons.
4 They have free electrons that can move easily.

Page 123
1 Circuit A.
2 The resistance increases.
3 $V = IR$ or $I = V \div R$ or $R = V \div I$
4 Because its resistance is not constant – it increases as more current flows through it making it hotter.

Page 124
1 It decreases.
2 Smoke detectors; automatic light controls; burglar alarms.
3 It decreases.
4 6 V (you *add* the p.d.s).

Page 125
1 12 A 2 1.5 V
3 Green and yellow striped.
4 The fuse melts if too much current flows through it, breaking the circuit and preventing harm.

Page 126
1 Alternating current.
2 A straight, horizontal line.
3 Anything above 0.5 A.
4 It breaks the circuit more quickly than a fuse.

Page 127
1 watts (W) 2 6.5 A
3 Proton.
4 5 protons; 6 neutrons; mass number 11.

Page 128
1 A form of an element that emits radiation.
2 a Positive; b negative; c no charge.
3 Use a Geiger counter.
4 Radon and thoron gas. Radon gas comes from granite; radon is radioactive.

Page 129
1 Rutherford and Marsden
2 Quite a lot of the alpha particles went straight through the gold foil, meaning that they must have gone through empty space rather than hitting anything.
3 The splitting of the nucleus of an atom.
4 A neutron.

Page 130
1 Uranium-235; plutonium-239.
2 In nuclear fission, the nucleus of an atom splits apart. In nuclear fusion, two atomic nuclei join together.
3 By nuclear fusion.
4 Helium.

Glossary/index

conduction Movement of energy as heat (or electricity) through a substance. 30, 49, 92

conductor A substance that will let heat or electricity pass through it, e.g. copper. 30, 32, 49, 119

consumer Organisms in an ecosystem that use organic matter produced by other organisms. 75

convection Movement of a heated substance which carries energy with it as heat. 49

convection current Upwards movement of heated gases or liquids to float on top of the cooler, denser layers. 43, 44, 49

core The centre of the Earth. 43

cosmic ray Radiation from space which hits the atmosphere. Some passes through, while some is blocked. 63, 128

count rate The number of nuclear events in a given time, often measured by a Geiger counter. 64

covalent bond A bond between atoms where electrons are shared. 27, 32, 92, 93, 95, 96

cracking The breakdown of long-chain hydrocarbon molecules into shorter-chain ones using heat, pressure and sometimes catalysts. 37, 38

crude oil A mixture of hundreds of different compounds, mainly hydrocarbons. 20, 34, 38, 39

crust The Earth's outermost layer of solid rock. 43, 44

current The flow of electrical charge through an electrical circuit. 54, 55, 119, 122, 123, 124, 125, 126

cutting, plant Part taken from a plant that can be used to grow a new, genetically identical plant. 17

cystic fibrosis An inherited disorder that affects cell membranes. Caused by a recessive allele. 88

cytoplasm The material in the cell that is inside the cell membrane but outside the nucleus. 70, 72

decomposer An organism that breaks down dead organic matter. 78

deforestation The permanent clearing of forests. 23

denature To alter the shape of a protein molecule, e.g. by heating, so that it can no longer do its job. 80

density The mass of an object divided by its volume. 31, 94, 115

deoxyribonucleic acid See DNA.

detergent A chemical that makes grease soluble in water. 81

detritus feeder An animal that eats the dead and semi-decayed remains of living things, e.g. an earthworm. 77, 78

diabetes A disease in which blood sugar levels are not controlled because the pancreas cannot secrete insulin, or because body cells do not respond to insulin. 7, 18, 84

diaphragm A sheet of muscle below the lungs that aids breathing. 82

differentiation A process in which new cells develop specialised features in order to carry out a job. 86

diffusion The spreading of gases or liquids caused by the random movement of their molecules. 5, 71, 72, 73, 82

digestion The breakdown of large complex chemicals into smaller simpler ones, e.g. starch into glucose. 5, 77, 80, 81, 83

digestive system The gut and the other body parts linked to it that help with digestion. 80

digital signal A pulsed (on/off) signal. 61, 62

distance-time graph A graph showing time along the x-axis and distance covered up the y-axis. 111

distillation A process used to boil off a liquid from a mixture and then cool the vapours to produce a pure liquid. 34, 40

DNA Deoxyribonucleic acid. The molecule that carries the genetic information in all living organisms. 16, 18, 65, 86, 88

DNA fingerprint The sequence of bases in a DNA molecule unique to each individual. 88

DNA profile DNA displayed on a sheet of film; each individual's DNA profile (fingerprint) is unique. 88

dominant A dominant allele is one which produces a feature even when another different allele of the same gene is present. 87, 88

double insulated An electrical device in which there are at least two layers of insulation between the user and the electrical wires. 125

drag A force which slows down something moving through a liquid or a gas. 115

dynamo A device that transforms kinetic energy into electrical energy. 55, 115

earthing To be connected to the earth – a safety feature of many electrical appliances. Prevents the build up of charge. 120, 125

earthquake Sudden movements of tectonic plates against each other, causing vibrations that travel through the Earth. 44

effector An organ that responds to a stimulus. 4, 5

efficiency The ratio of useful energy output to the total energy input. 52, 56, 57, 76

egg A special cell (gamete) made by a female for sexual reproduction. 6, 16, 17, 85, 87

electrical charge Electrical energy stored on the surface of a material, also known as a static charge. 50, 53, 54, 55, 56, 57

electrode Bar of metal or carbon that carries electric current into a liquid. 106

electroluminescent A material that emits light of different colours when an alternating current passes through it. 96

electrolysis Using an electric current to split a compound, either in solution or in its molten state. 31, 106, 107

electromagnetic radiation Energy carried by a wave, which can pass through a vacuum. 48, 59, 63

electromagnetic spectrum The range of all the different types of electromagnetic waves in order of their energies. 59, 60, 65

electron A small negatively-charged particle that orbits around the nucleus of an atom. 26, 27, 30, 63, 65, 68, 90, 91, 92, 94, 118, 119, 122, 123, 124, 127, 128

electron notation Representation of the arrangement of electrons around an atom of an element as a series of numbers, e.g. magnesium, 2, 8, 2. 90

electroscope A device that detects if an object has an electrical charge (gold leaf electroscope). 119

element A substance that cannot be split into anything simpler by chemical means. 26, 33, 62, 90, 91, 93, 96, 97, 106

embryo An organism in the earliest stages of development, for example an unborn human baby or a young plant inside a seed. 6, 17, 18, 86, 88

embryo transplantation To move an embryo from one mother, or laboratory vessel, to the uterus of another. 6, 17

emphysema Breakdown of the air sacs in the lungs causing a shortage of oxygen in the blood and breathlessness. 11

emulsifier A chemical which can help to break fats down into small globules. 41, 80

emulsion A mixture with one liquid dispersed in another. 41

endothermic reactions Reactions in which energy is taken in. 104

energy The ability of a system to do something (work). We detect energy by the effect it has on the things around us, e.g. heating them up. 7, 48, 49, 50, 51, 52, 53, 55, 59, 60, 62, 63, 65, 66, 67, 68, 70, 76, 78, 82, 90, 104

energy transfer When the same form of energy moves from one place to another. 48, 49, 50, 51, 54, 61, 75, 115, 116

energy transformation When one form of energy is changed to another. 50, 51, 56, 115

environment The conditions in which a plant or animal has to survive; everything that affects a living organism. 15, 20, 39, 96

enzyme A biological catalyst that speeds up the rate of chemical reactions. 5, 18, 77, 80, 81, 86

epidemic An infectious disease affecting many people in a particular area. 12

equilibrium The point at which a forward reaction in a reversible reaction balances the reverse reaction. 105

evolution The gradual change in living organisms over millions of years caused by random mutations and natural selection. 18, 19

exothermic reaction Reactions in which energy is given out. 104, 105

extinction When all members of a species have died. 20, 23

fat Chemical made up of glycerol and fatty acids. It is used as a source of energy and to make cell membranes. 7, 8, 41, 73, 74

fertilisation When male and female sex cells (gametes) fuse together. 16, 85

fertiliser A substance added to the soil to help plants to grow. 21, 105, 109

fertility drug A drug used to make eggs mature in the ovary. 6

food chain A diagram using arrows to show how energy passes from one organism to another. 75, 76, 77

food web A diagram showing how all of the food chains in an area link together. 75

force A push or pull which is able to change the velocity or shape of a body. 52, 112, 113, 116, 118, 119, 122

fossil Preserved evidence of a dead animal or plant. 19

fossil fuel A fuel such as coal, oil and natural gas formed by the decay of dead organisms over millions of years. 22, 24, 35, 40, 46, 56, 57

fractional distillation The separation of a mixture into its many components using their different boiling points. 34

free electron An electron that is free to move between atoms in a substance. 49, 122

frequency The number of waves passing a point in one second, measured in hertz (Hz). 59, 126

friction The force that acts between two surfaces in contact with each other, resisting movement. 113, 114, 115, 116, 120

fuel A substance that gives out energy, usually as light and heat, when it burns. 38, 40

fuse A component in an electrical circuit containing a thin wire which heats up and melts if too much current flows through it, breaking the circuit. 54, 122, 125, 126, 127

gall bladder An organ associated with the liver that stores bile. 80

gamete Sex cell (egg or sperm) that joins to form a new individual during sexual reproduction. 16, 85, 88

gamma radiation Gamma rays emitted from a radioactive isotope. 64, 65, 128

gamma rays High-energy, high-frequency radiation travelling at the speed of light. 60, 61, 62, 63, 65

Geiger counter A device used to detect some types of radiation. 64, 128

gene A length of DNA on a chromosome in a cell that carries information about how to make particular proteins. 16, 18, 19, 85, 87

generator A device for converting kinetic energy into electrical energy (current flow). 55, 56

genetic engineering A range of technologies that allow scientists to manipulate individual genes. 18

genetic modification Altering the genetic make-up of an organism by transferring genes from one organism to another using DNA technology. 18

negative ion An ion with a negative charge. 91, 92, 106

nephrons Tiny tubes in the kidneys that collect waste substances, including urea and water, from the blood. 83

nervous system The central nervous system and nerves that control the actions and functions of the body. 4

neurone A nerve cell, specialised for conducting electrical impulses. 4, 5

neutral A solution with a pH of 7 that is neither acidic nor alkaline. 108

neutralisation A reaction between an acid and an alkali to produce a neutral solution of a salt and water. 22, 104, 108, 109

neutron A particle found in the nucleus of an atom; it has no electrical charge and a mass of 1 atomic mass unit. 26, 62, 68, 90, 91, 127, 128, 129, 130

newton The unit of force (N). 113

nicotine An addictive drug found in tobacco. 11

nitrate A salt of nitric acid. 74, 82, 105

nitrogen A non-reactive gas that makes up most of the atmosphere. 27, 33, 45, 74, 99, 104, 105

non-renewable A resource that is being used up faster than it can be replaced, e.g. fossil fuels. 20, 33, 38, 39, 40, 56

nuclear fission The splitting of an atom's nucleus to release nuclear energy. 56, 129

nuclear fusion When nuclei of atoms fuse together giving out huge amounts of energy, e.g. when a star is born. 66, 67, 68, 130

nuclear power The energy obtained by transforming the energy from nuclear fission to electrical energy. 56, 57, 129, 130

nuclear radiation The energy released when the nucleus of an atom undergoes radioactive decay.

nucleus: biology The control centre of the cell, the nucleus is surrounded by a membrane that separates it from the rest of the cell. 16, 17, 86

nucleus: chemistry The central part of an atom containing the protons and neutrons. 26, 62, 63, 64, 90, 127

oestrogen A hormone produced in females by the ovary. 4, 6

Ohm's Law Current is directly proportional to potential difference across the resistor. 123, 124

optical fibre A very fine strand of silica glass, through which light passes by successive internal reflections. 62

optical telescope An instrument that uses light to form images of objects far away, e.g. in space. 65

optimum temperature The temperature range that produces the best reaction rate. 80

orbit A path, usually circular, of a smaller object around a larger object, e.g. a planet orbits the Sun; electrons orbit the nucleus in an atom. 66

organelle An internal part of a cell, e.g. ribosome, mitochondrion. 70, 82

oscilloscope A device that displays a line on a screen showing regular changes (oscillations) in something. An oscilloscope is often used to look at sound waves collected by a microphone. 126

osmosis The diffusion of water molecules from a dilute solution to a more concentrated solution through a partially permeable membrane. 72

ovary Female reproductive organ. 4, 6, 85

oxidation Loss of electrons from an atom. 104, 106

ozone A substance in the atmosphere formed from oxygen (O_3), which acts as a shield against ultraviolet radiation. 45

pancreas An organ in the abdomen that produces enzymes to break down food and the hormone insulin to reduce blood sugar level. 4, 5, 80, 84

pandemic A world-wide epidemic. 12, 13

Pangaea The supercontinent thought to be the first landmass to form. 44

parallel circuit An electrical circuit in which the current divides to take different routes then joins up again. 122, 125

partially permeable membrane A membrane through which only some molecules can pass. 72

pathogen An organism that causes a disease. 11, 12, 13

payback time How long it takes for savings to equal costs. 53

periodic table A chart grouping the elements according to their similarities, first devised by Dimitri Mendeleev. 26, 90, 91

pesticide A chemical used to kill a pest. 21

pH The level of acidity or alkalinity of a substance measured as a scale from 1 to 14: neutral, 7; acidic, below 7; alkaline, above 7. 80, 108

phagocyte A white blood cell that ingests bacteria in a process called phagocytosis. 12

photochromic A material that darkens on exposure to sunlight. 96

photosynthesis The production, in green plants, of glucose and oxygen from carbon dioxide and water using light as an external energy source. 21, 22, 23, 45, 46, 73, 74, 78, 80

pollutant A chemical that causes pollution. 31

pollution Human activity that causes damage to the environment, e.g. dumping raw sewage at sea. 20, 21, 22, 23, 33, 39

pollution indicator An organism that is sensitive to a particular level of environmental pollution. The presence or absence of a particular pollution indicator can be used to assess the degree of pollution in an environment. 23

polymer A molecule made of many repeating subunits, e.g. polythene or starch. 37, 38, 39

polymerisation The process of forming large polymers from smaller monomer molecules. 38

polyunsaturates Fats with more than one carbon-carbon double bond. 41

population All the organisms of the same species that live in a particular place. 20, 21

positive ion An ion with a positive charge. 91, 92, 94, 106

potential difference The difference between two points which will cause current to flow in a closed circuit, measured in volts (V). 119, 122, 123, 124, 125, 126

potential energy Stored energy in chemicals or objects. 50, 53, 115, 116

power The rate that a system transfers energy, i.e. does work, usually measured in watts (W). 52, 55

power rating The rate at which an appliance transforms electrical energy. 54, 127

power station A place that generates electricity. 22, 35, 56, 57

precipitate To fall out of solution. A insoluble solid. 109

predator An animal that hunts and kills other animals. 15, 20, 75

pressure The force acting on a surface divided by the area of the surface, measured in newtons per square metre (N/m^2). 102

prey Animals that are hunted by other animals. 75

producer An organism that makes organic material from inorganic substances. Green plants are primary producers because they use energy in sunlight to make sugar. 75

product Something made by a chemical reaction. 98, 99, 101, 104

protease An enzyme that digests proteins. 81

protein A complex molecule that contains carbon, hydrogen, oxygen, nitrogen, and sometimes sulfur. It is made of one or more chains of amino acids. It is used for growth, repair and energy. 7, 73, 74, 80, 81, 83, 86

protein synthesis To make protein in a cell from amino acids. 70, 80

proton A particle found in the nucleus of an atom with a charge of +1 and a mass of 1 atomic mass unit. 26, 62, 63, 68, 90, 91, 127, 128, 130

pyramid of biomass A diagram to show the masses of living organisms present at each step in a food chain. 75

radiation Energy that travels as light or other forms of electromagnetic waves. 59, 60, 61, 62, 63, 64, 65, 127, 128

radio wave A form of electromagnetic radiation used to carry radio signals. 50, 60, 65

radioactive Material which gives out radiation. 56

radioactive decay The breakdown of an unstable nucleus of an atom, releasing energy. 43, 62, 64, 128

radioactive waste Waste produced by radioactive materials used at power stations, research centres and some hospitals. 65

radioisotope A radioactive isotope. 62, 63, 64

radiotherapy Using radiation (gamma rays) to treat certain types of disease, e.g. cancer. 64

radon A naturally occurring radioactive gas, found near granite rocks. 63, 128

reactant A chemical taking part in a chemical reaction. 98

reaction time: biology/physics The time taken to respond to a stimulus. 114

reaction time: chemistry The time taken for a reaction to finish. 101, 103, 104, 105

reactivity A measure of how easily a chemical will react. 32, 37, 38, 94, 95, 107

reactivity series A series of metals in order of their ability to react with water and acids. 107

recessive A recessive allele is one which produces a feature only when the dominant allele of the same gene is not present. 87, 88

receptor A special cell that detects a stimulus. 4, 5

reduction Gain of electrons by an atom. 106

reflex action A rapid automatic response to a stimulus. 5

reflex arc The route of the electrical impulse in a reflex action along a sensory, a relay and a motor neurone. 5

relative atomic mass The mass of an atom (A_r) as the sum of its protons and neutrons (same as its mass number). 91, 97

relative formula mass The mass of a molecule (M_r) in comparison to the mass of hydrogen, which is taken as 1. 97

relative mass The mass of a particle (proton, neutron or electron) in an atom relative to one another. 90

renewable A resource that can be replaced, e.g. wind power, wave power and solar power are all renewable sources of energy. 33, 38, 40, 56, 57

resistance: biology When an organism's genetic make-up changes so that it is unaffected by an agent used to destroy it. 12, 13, 18

resistance: physics The amount by which a conductor prevents the flow of electric current. 54, 122, 123, 124

resistor A conductor that reduces the flow of electric current. 122, 123, 127

respiration The chemical process that releases energy from food in all cells. All living things must respire. 23, 70, 71, 73, 75, 76, 77, 78, 80, 82

resultant force The sum of all the different forces acting on an object. 112, 114

reversible reaction A reaction whose direction can be changed by a change in conditions. 96, 98, 99, 104, 105

ribosome An organelle in the cell in which protein synthesis takes place. 70, 98

salt A compound made when an acid reacts with an alkali. Common salt is sodium chloride. 8, 74, 94, 95, 107, 108, 109

Sankey diagram A diagram that shows the input and output of energy in a device. 51, 52

satellite A body orbiting around a larger body. Communications satellites orbit the Earth to relay messages from mobile phones. 61, 65

saturated Fats or oils containing only single bonds between carbon atoms. 8, 34, 37, 41

secretion The production of useful substances by organs or cells, e.g. hormones by glands. 4, 5, 81

sedimentation Particles settling out of suspension in water. 19

series circuit An electrical circuit where components are connected in one continuous loop. 122, 124

sewage Waste matter and rubbish carried in sewers. 21, 23

sex chromosomes Chromosomes that carry the genes that determine sex. 87

sexual reproduction Reproduction involving gametes and fertilisation. There are usually two parents each contributing half the genes of the offspring. 16

shape memory When a smart alloy that can be deformed is able to return to its original shape when heated. 33, 96

shell A grouping of electrons in an energy level around an atom. 26, 27, 30, 90, 92, 94, 95

silicon dioxide Sand. Covalently bonded silicon and oxygen atoms. 93

skin graft Surgical replacement of damaged skin by undamaged skin. 17

smart alloy An alloy that has a shape memory. 33

smart material A material that responds to a change in the environment by changing in some way that is reversible. 96

smog Smoke particles plus water droplets. 22

smoke Tiny particles (e.g. carbon particles) suspended in air, produced by burning materials. 22, 64

solar cell A device that converts light energy into electrical energy. 57

solar energy Energy from the Sun. 53, 57, 75

Solar System The collection of planets and other objects orbiting around the Sun. 43, 66, 67

species A group of living things of the same type. Humans belong to the species *Homo sapiens*. 18, 19, 20

spectroscope A device that splits light into bands of different colours. 68

spectral pattern The pattern formed when an element is heated and emits light, as viewed by a spectroscope. 68

speed How fast an object is moving. 101, 111, 112, 113, 114, 115, 116

sperm A special cell (gamete) made by a male for sexual reproduction. 16, 70, 85, 87

star A large gas body in space that emits light from internal nuclear reactions. 66, 67, 130

static electricity Electric charge on the surface of a non-conductor, i.e. it is charge that cannot flow easily. 102, 118, 119

stem cell An undifferentiated cell found in embryos and certain parts of adults that can develop into many other types of cell. 86

stimulus A change inside or outside of an organism that brings about a response, e.g. altered blood sugar levels; a flash of light. 4, 5

stoma Tiny hole in the surface of a leaf which allows gases to diffuse in and out. 73

stopping distance The distance required to bring a moving vehicle to a halt. The sum of the thinking distance and the braking distance. 114

subduction zone A region where one tectonic plate moves under another. 44

substrate Something that an enzyme causes to react to make a product. 88

surface area The area of a surface. Surface area has a significant effect on the rate of many chemical reactions. 15, 82, 95, 102, 103

sustainable development Development that does not make life more difficult for future generations. 24, 98, 105

synapse The gap between the ends of two neurones. 5

target organ An organ affected by the presence of a particular hormone, e.g. adrenaline from the adrenal gland affects the heart. 4

tectonic plate Cracked sections of the lithosphere that ride on top of the fluid mantle. 44

telecommunications Communication over a distance by cable, telephone or broadcasting. 61

temperature A measure of how hot something is. 46, 48, 71, 74, 77, 80, 83, 84, 101, 102, 123, 124

temperature receptor Cell in the brain that monitors blood temperature. 84

testis Male reproductive organ. 4, 85

testosterone A male hormone made by the testes. 4

thermal decomposition Breaking down a chemical using heat. 28, 104

thermal energy *See* heat.

thermal radiation The transfer of heat energy by rays. Also called infrared radiation. 48

thermochromic Material that changes colour when it gets hotter or colder. 96

thermopile An instrument that measures temperature. 48

thermoregulatory centre An area of the brain containing temperature receptors that monitors blood temperature. 84

thinking distance The distance travelled by a moving vehicle before the brakes are applied. The product of speed and reaction time. 114

tidal power The energy obtained by transforming the energy in the rise and fall of the tides into electrical energy. 56, 57

tissue A group of cells of the same type, e.g. nervous tissue contains nerve cells. 70, 73

total internal reflection When light hits a glass/air interface at an angle of more than 42°, it is reflected back at the boundary. 62

toxic chemical A chemical that harms living things, a poison. 9, 21, 42

toxin A poison. Usually used to mean a poison produced by a living organism. 11

transformer A device that alters the voltage of an alternating current. 55, 56

transition element A metal belonging to the transition group in the periodic table. 30

turbine A device that transfers kinetic energy into circular movement, usually to drive a generator. 56, 57, 130

ultraviolet radiation Radiation just beyond the blue end of the spectrum of visible light. UV light causes tanning and some sorts of skin cancer. 60, 61

Universe Everything, everywhere. 68

upthrust An upward force, e.g. of water on a floating object. 113

uranium A radioactive metal used in nuclear power stations and bombs. 129, 130

urea A substance made in the breakdown of proteins in the liver. 5, 83

urine A solution of urea and other wastes in water, made in the kidneys for excretion by the bladder. 5, 83

uterus The organ in the female where the baby grows during pregnancy; also known as the womb. 6

vaccine A chemical developed to immunise people against a particular disease. 13

vaccination Injection of a vaccine into the bloodstream. 13

vacuum A space with no particles in it. 48, 59

Van de Graaff generator A device used to generate static electrical charges. 119

variation The existence of a range of individuals of the same group with different characteristics. 16, 19, 23, 85

vegetable oils Oils obtained from the fruits, seeds and nuts of some plants, e.g. olive, rape, coconut. 40

velocity The speed of an object in a given direction (*v*). 111, 113, 114, 115, 117, 118

virus A microorganism (a thousand times smaller than a bacterium) that invades cells, reproduces and bursts cells, making people ill. 11, 12, 13

vitamin A substance found in foods that is essential to maintain health. 40

volcano The upthrust of the crust of the Earth at a point of weakness, allowing magma from the mantle to be forced out. 44

volt The international unit of electrical potential (V). 126

voltage The potential difference across a component or circuit (V). 55, 56, 94, 120

voltmeter A meter that measures the voltage between two points. 122, 124, 125

watt A unit of power (W), 1 watt equals 1 joule of energy being transferred per second. 52, 54, 127

wave power The energy obtained from transforming the energy of waves into electrical energy. 56, 57

wavelength The distance between two identical points on a wave. 59, 60, 61, 68, 75

weight The force of gravity acting on a body measured in newtons (N). 52, 114

wind farm A group of wind turbines used to generate electricity from the kinetic energy in moving air. 56

withdrawal symptoms The combination of physical and psychological symptoms produced when an addictive drug is withheld for a period of time. 9, 10

work Work is done when a force moves. The greater the force or the larger the distance, the more work is done. 52, 115, 116

X-ray Electromagnetic radiation used for imaging or to destroy some types of cancer cells. 60, 61, 88

XX chromosomes The sex chromosomes present in a human female. 87

XY chromosomes The sex chromosomes present in a human male. 87

yeast A unicellular fungus used extensively in the brewing and baking industries. 38

yield The ratio of product to starting materials; a high yield means that most of the starting material is converted to useful products. 98, 105

zygote A fertilised egg, produced when a male and female gamete join. 17